Analytics

Data Science, Data Analysis, and Predictive Analytics for Business

5th Edition

By
Daniel Covington

© Copyright 2016 - All rights reserved.

In no way is it legal to reproduce, duplicate, or transmit any part of this document in either electronic means or in printed format. Recording of this publication is strictly prohibited, and any storage of this document is not allowed unless with written permission from the publisher. All rights reserved.

The information provided herein is stated to be truthful and consistent, in that any liability, in terms of inattention or otherwise, by any usage or abuse of any policies, processes, or directions contained within is the solitary and utter responsibility of the recipient reader. Under no circumstances will any legal responsibility or blame be held against the publisher for any reparation, damages, or monetary loss due to the information herein, either directly or indirectly.

Respective authors own all copyrights not held by the publisher.

Legal Notice:

This book is copyright protected. This is only for personal use. You cannot amend, distribute, sell, use, quote or paraphrase any part or the content within this book without the consent of the author or copyright owner. Legal action will be pursued if this is breached.

Disclaimer Notice:

Please note the information contained within this document is for educational and entertainment purposes only. Every attempt has been made to provide accurate, up-to-date, reliable, and complete information. No warranties of any kind are expressed or implied. Readers acknowledge that the author is not engaging in the rendering of legal, financial, medical or professional advice.

By reading this document, the reader agrees that under no circumstances are we responsible for any losses, direct or indirect, which are incurred as a result of the use of information contained within this document, including, but not limited to, errors, omissions, or inaccuracies.

Table of Contents

Introduction

What defines the success of a business? Is it the number of people employed by the firm? Is it the sales turnover of the business? Is it the strength of the customer base of the firm? Does employee satisfaction play a role in the success of business operations? How does management factor into the overall operational success? How critical is the role of a data scientist in this process? Finally, does fiscal responsibility play any role in the success of any business?

To answer any of these questions, it is important that you have the required data in hand. For instance, you initially need to know how many staff members you have employed to assess the value they contribute to the growth of your business. Similarly, you need to have a repository of all the customers, along with details of their transactions, to understand whether your customer base is contributing to the success of your firm.

Why is data important? Data is important for the sustenance of business. The preliminary reason why it is important is because you need information to be aware of you're the state of affairs in which your business operates. For instance, if you do not know how many units your company sells in a month, you will never be able to determine whether your business is doing well. There are several other reasons as to why data is important. I have dealt with these reasons in detail in the upcoming chapters of this book that focus on the importance of data.

Just collecting data is not enough. Analyzing and putting data to use are also important. Of course, if you belong to the class of people who are not worried about whether their business lost a customer or not, you do not have to spend time on analyzing data. However, if this attitude persists, you will soon see the end of your business because of the growing number of competitors who care about the expectations of their customers. This is where predictive analytics comes into play. How you employ predictive analytics in your firm is what distinguishes your business from the others in the market. Predictive analytics has the capacity to change the way you play your game in the market. It is capable of giving you that little edge over your competitor.

In the first chapter of this book, I highlight the importance of data in business. I also focus on how data plays an important role in increasing the efficiency of a business. In upcoming chapters I go over the different steps involved in the process of data analysis, as well as take you through the basics of predictive analytics and the various methods involved. I discuss the different techniques used for conducting predictive analytics, and you will see how predictive analytics is being employed in different fields. You will truly appreciate its flexibility, seeing how it can be used in finance as well as in medicine, and in operations as well as in marketing.

Particular chapters of the book will cover the use of big data analysis and its application to the marketing, gaming, and retail industries, as well as the public sector. I have examined the various ways that big data can benefit both major private businesses and public institutions, such as hospitals and law enforcement agencies, and bring both increased revenues to companies and a safer and healthier living climate to our cities.

In one short chapter, I have covered the most basic form of

data analysis, which is descriptive analysis. This field, while basic, is a necessary prerequisite to all other forms of analysis, such as predictive analysis, as we cannot make any predictions without first examining the already available data. This is where descriptive analysis comes into play to provide us with a basis for future inferential and predictive analysis.

The fields of data analysis and predictive analytics are so vast and have so many sub-branches, which are even more extensive. One of these branches is prescriptive analysis, which I will briefly cover in the last chapter. I have covered only the fundamentals of these fields in this book. This method is being used by large industries to determine what will happen in the future, and they use the information to prevent or make things happen in the future.

I hope you truly enjoy this book. I am sure you will be motivated to manage your data better and employ predictive analytics in your business to reap the maximum benefits. Thank you for purchasing this book.

Chapter 1: The Importance of Data in Business

Were you ever involved or interested in the study of the classic languages, particularly those that are otherwise known as "dead" languages? While the sheer complexity of these languages is fascinating, the best part about them – the part that will simply blow your mind – is that the ancients used some extremely tedious methods to preserve sacred texts that range from just a few hundred years old to several thousand. To preserve these texts, scribes would painstakingly copy them, not just once but several times, and this could take years to complete. They copied the texts onto papyrus paper and used ink made from oil, burnt wood, and water. Another common method was to chisel the characters onto shards of pottery or onto stone. While these were exceedingly time-intensive and, I have no doubt, extremely mind-numbing at times, the information that has been preserved was deemed to be so invaluable to the ancient civilizations that certain people dedicated their whole lives to doing it.

So, I can hear you asking, what on earth do ancient texts and dead languages have to do with data analysis and business analytics? Data – everything is about the data; it is something that is all around us, and we simply can't get enough of it. Think about today; think about social media platforms that are revolutionizing the landscape for marketing because they are providing companies with a whole set of analytics that allows them to measure how successful, or not as the case may be, their company content is, and many of these platforms provide these analytics tools for free.

On the other hand, there are platforms that charge a high price to provide you with high-quality data telling you what does and doesn't work on your website. In the world of business, market and product data give a business the edge over competitors, and that makes such data worth its weight in gold. Important data includes historical events, weather, products, trends, customer tendencies, outliers, and anything else that is relevant to business verticals.

What has changed is the way that data is stored. It is no longer a long and cumbersome task; it is automatic, requires little in the way of human intervention, and it is done on a huge scale. Today's modern day scribes are connected sensors.

If you think about it, if you were to collect every single piece of information that you could – every piece of data that society generated – it would be nothing compared to what you will face in the next few years. This is the Internet of Things. This is because most of our devices are connected; they record performance and usage data, and they transmit that data. Sensors record environmental data. Cities are completely connected to ensure traffic and infrastructures are running at the top level. Delivery vehicles are connected, so their location is monitored, along with their efficiency, even to the extent that a problem with the vehicle can be identified early. Buildings and homes are connected to improve the cost and control of energy, while manufacturing premises are connected so that the communication of critical data is automatic. This is the present, and ever more so, it is the future. We are connected; everything is connected to the internet.

The fact that data is important really isn't anything new. It's just that we have moved past a scribe and a chisel to the use of microprocessors. The way we capture data and the type of data we capture are ever changing, and it is vital that you stay up to date and ahead of the game if you want to be in the game to win.

Even though it is the service provided or the goods manufactured by a company that helps it establish a niche for itself in the market, data plays a crucial role in sustaining success. In today's technology-driven world, information can make or break a business. For instance, there are businesses that have disappeared in such a short time because they failed to gauge their progress or customer base. Meanwhile, we also have startups that have been doing extremely well because of the increased importance they show towards numbers and the expectations of their customer base.

What is the Source of Data?

Data could refer to the sales figures or the feedback from the customers or the demand for the product or service. Some of the sources of data for a company are as follows:

- **Transactional Data:**

  This could be pulled out from your ledger, sales report, and web payment transactions. If you have a customer relationship management system in place, you will also be able to take stock of how your customers are spending on your products.

- **Online Engagement Reporting:**

  This is data based on the interaction of customers on your website. There are tools available, such as Crazy egg and Google Analytics, which can help you collect data from your website.

- **Social media:**

  Social networking sites, such as Twitter, Facebook, and LinkedIn, provide insights into the customer traffic on your page. You can also use these platforms to conduct a cost-effective survey on the tastes and preferences of your customers and to improve products or services.

How Can Data Improve Your Business?

Data can improve the efficiency of your business in many ways. Here is a taste of how data can play an important role in upping your game.

- **Improving Your Marketing Strategies:**

  Based on the data collected, it is easier for a company to come up with innovative and attractive marketing strategies. It is easier for a company to alter existing marketing strategies and policies in such a fashion that they are in line with the current trends and customer expectations.

- **Identifying Pain Points**

If your business is driven by predetermined processes and patterns, then data can help you identify any deviations from the usual. These small deviations could be the reason behind the sudden decrease in sales, increase in customer complaints, or decrease in productivity. With the help of data, you will be able to catch these little mishaps early and take corrective actions.

- **Detecting Fraud**

When you have the numbers in hand, it will be easier for you to detect any fraud that is being committed. For instance, when you have the purchase invoice of 100 units of tins and then you see from your sales reports that only 90 tins have been sold, then you are missing ten tins from your inventory, and you know where to look. Most companies are being silent victims of fraud because they are unaware of the fraud being committed in the first place. One important reason for this is the absence of proper data management, which could have helped detect fraud easily in the early stages.

- **Identify Data Breaches**

The explosion of complex data streams in the past few years has brought on a new set of problems in the area of fraudulent practices. They have become subtle and comprehensive. Their negative effects can be widespread and can impact your company's retail, accounting, payroll, and other business systems adversely. In other words, data hackers will become more devious in their attack on company data systems.

By using data analytics and triggers, your company can prevent fraudulent data compromises in your systems, which can severely cripple your business. Data analytics tools enable your company to develop data testing processes to detect early signs of fraudulent activity in your data systems. Standard fraud testing may not be feasible in certain circumstances. If this is the case with your company, then special tailored tests can be developed and used to trace any possible fraudulent activity in your data processes.

Traditionally, companies have waited until their operations have been impacted financially to investigate fraud and implement breach prevention strategies. This is no longer feasible in today's rapidly changing data-saturated world. With information being disseminated worldwide so quickly, undetected fraudulent activity can cripple a company and its subsidiaries in no time globally.

Conversely, data analytics testing can stop potential fraudulent data destruction by revealing indicators that fraud has begun to seep into data systems. Fraud can be stopped quickly for a company and its partners worldwide if these data analytic tests are applied periodically.

- **Improving Customer Experience**

As I previously mentioned, data also includes the feedback provided by customers. Based on their feedback, you will be able to work on areas that can help you improve the quality of your product or service and therefore satisfy the customer. Similarly, when you have a repository of customer feedback, you will be able to customize your product or service in a better fashion. For instance, there are companies that send out customized personal emails to their customers. This sends out a message that the company genuinely cares about its customers and would like to satisfy them. This is possible solely because of effective data management.

- **Decision Making**

Data is crucial for making important business decisions. For instance, if you want to launch a new product in the market, it is important that you first collect data about the current trends in the market, the size of the consumer base, the pricing of the competitors, etc. If the decisions made by a company are not driven by data, then it could cost the company a lot. For instance, if your company decides to launch a product without considering the price of the competitor's product, then there is a possibility that your product might be overpriced. As is the case with most overpriced products, the company would have trouble increasing the sales figures.

When I say decisions, I do not really just refer to the decisions pertaining to the product or service offered by the company. Data can also be useful in making decisions with respect to the function of departments, manpower management, etc. For instance, data can help you assess the number of personnel required for the effective functioning of a department, in line with the business requirements. This information can help you decide whether a certain department is overstaffed or understaffed.

- **Hiring Process**

Using data in selecting the right personnel seems to be a neglected practice in corporate life. It's critical to position the most qualified person in the right job in your company. You want your business to be highly successful in every facet of operation. Using data to hire the right person is a sure way to put the best person in the job. What kind of data would you use to hire a professional?

Big companies, which have astronomical budgets, use big data to locate and select the most skilled individuals for the right jobs. Start-ups and small companies would benefit immensely from using big data to hire the right group of people to make their recruitment successful from the start. This avenue for gathering data for hiring purposes has proven to be a successful avenue for hiring the right fit for organizations of all sizes. Again, companies can use their data scientists to extract and interpret the precise data needed by human resource departments.

- **Using Social Media Platforms to Recruit**

Social media platforms (Facebook, Twitter, and LinkedIn to name a few) are hidden gold mines as data sources for finding high-profile candidates for the right positions within companies. Take Twitter, for instance; company job recruiters can follow people who tweet specifically about their industry. Through this process, a company can find and recruit the ideal candidates based on their knowledge of a particular industry or job within that industry. Do their "tweets" inspire new thoughts and possibly new innovations for their industry? If so, you have a whole pool of potential job applicants.

Facebook is another option for data gathering for potential job candidates. Remember, these avenues are virtually free, and corporations could use them as part of a major cost-effective strategy. Facebook is all about collecting social networking data for companies looking to expand their workforce or replace an existing open position(s). Company recruiters can join industry niche groups or niche job groups. "Liking" and following group member's comments will establish the company's presence within the group, thus allowing highly focused job ads to be posted within the group. The company can increase views, thereby widening the pool of potential job candidates.

It is easy to establish a timeline to brand the company as an innovative and cutting-edge place to work. You establish your presence by engaging with friends/followers who are in the same industry as your company. For a minimal advertising fee, you can promote your job ad. By doing this, you geometrically increase your reach among potential job seekers. If your company releases highly effective job data posts, you will reel in a higher yield of highly skilled job searchers, therefore greatly increasing the percentages of people who are the perfect fit for your job.

- **Niche Social Groups**

Niche socials groups are specialized groups that you can join on social and web platforms that can help you find specific skillsets. For example, if you are looking to hire a human resources manager, what better place to find a potential recruit than by joining a specific human resources social group? Locate social connections within that group and then post descriptive but alluring job data posts, and you may find the right person to fit into your human resources position. Even if your company doesn't find the right person, group members will surely have referrals. Again, approaching these groups is a very cost-effective way to advertise your job posts.

- **Innovative Data Gathering Methods for the Hiring Process**

Why not think outside the hiring process box and try new methods of data collection to hire the right professional? Use social collecting data sites that gather data, such as Facebook, Google+, LinkedIn, and Twitter. Your company can search on these sites, extracting pertinent data from posts made by potential job candidates. Such data can be used to help your company connect with highly efficient job applicants.

An overlooked but very good data pool to use is keywords. Keywords are used on the internet for every type of search imaginable. Why not use the most visible keywords in your online job descriptions? By doing this, your company can widely increase the number of views that your job posting will attract.

You can also use computers and software to find the right job candidate for your company. Traditionally, these data sources have been used to either terminate a company's employee or analyze whether an existing employee is a right fit for another job within the company.

Why not try a whole new data collecting system? This system would be rooted in a set of standards different from the usual IQ tests, skill testing, or physical exams. These are still valuable tools to measure candidates by, but they are limiting. Another focus can be the strong personality traits a candidate may possess. Is the person negative? Is the person argumentative? Is the person an isolationist who doesn't get along with other people? These types of people can be identified through this personality trait database and then filtered out as possible team members. This type of data, when properly extrapolated, will save the company time, resources, and training materials. By eliminating a mismatch between the expectations the company has for the job and the person who would potentially fill the job.

Another benefit of this data gathering system is that the results will not only identify skilled people for the right jobs but also people with the right personality to fit in with the current company culture. It's imperative that a person is sociable and will be able to engage other employees to produce the most effective working relationships. The healthier the working environment, the more effective company production is, overall.

- **Gamification**

This is a unique data tool that isn't currently in widespread use. It does motivate candidates to press in and put forth their best effort in the job selection process. You provide people with "badges" and other virtual goods that will motivate them to persevere through the process. In turn, their skills in being able to perform the job requirements will be readily obvious. This also makes the job application a fun experience instead of a typical tedious task.

- **Job Previews**

Pre-planning the job hiring process with accurate data about what the job requirements are will prepare the job seeker to know what to expect if he or she is hired for the position. It is said that a lot of learning on the job is by trial and error, which prolongs the learning process. This takes more time to get the employee up to speed to function efficiently as a valuable resource within the company. Incorporating the job preview data into the hiring process reduces the learning curve and helps the employee become efficient much quicker.

These are some innovative data gathering methods companies can use to streamline the hiring process. They also help human resource departments pick the most skilled people to fill their employment needs.

Hence, data is crucial in aiding businesses to make effective decisions.

These are some of the reasons why data is crucial for the effective functioning of a business. Now that we have had a glance at the importance of data, let us get into the other aspects of data analysis in the upcoming chapters.

Chapter 2: Big Data

Data is firmly woven into the fabric of society, across the entire globe. Like every other important production factor – like human capital and hard assets – much of our modern economic activity could not happen without data. Big data is, in a nutshell, large amounts of data that can be gathered up and analyzed to determine whether any patterns emerge and to make better decisions. In the very near future, big data will become the base on which companies compete and grow and on which productivity will be strongly enhanced and significant value will be created for the global economy by increasing the quality of services and products while reducing the amount of waste.

Until now, the river of data that has been flooding the world was something that most likely only grabbed the excitement of some data geeks. Now, we are all excited by it simply because the sheer amount of data that is generated, mined, and stored has become one of the most relevant factors, in an economic sense, for consumers, governments, and businesses alike.

Looking back, we can see now that trends in IT innovation and investment, with their impact on productivity and competitiveness, suggest that big data has the ability to make sweeping changes to our lives. Big data has the same preconditions that allowed previous IT-enabled innovation to power productivity (for example, innovations in technology, followed closely by complementary management innovations being adopted). We expect that big data technology suppliers and analytic capabilities that are now far advanced will have as much, if not more, impact on productivity as do suppliers of different technologies.

Every business in the world needs to take big data very seriously as it has a huge potential to create real value. Some retail companies that are wholly embracing big data are seeing the potential for a large increase in operating margins

Big Data – The New Advantage

Big data is fast becoming the most important way for companies at the top to seriously outperform their competition. In most industries, both new entrants to the market and those that are firmly established will leverage strategies that are driven by data to compete, innovate, and capture real value. Examples of this exist everywhere. Take the healthcare industry, for example; data pioneers are examining the outcomes of certain pharmaceuticals that are widely prescribed. In their analysis, they discovered that there were risks and benefits that were not seen in limited trials.

Other industries that have taken on big data and run with it are using sensors that are embedded into products to gain data. These products range from a child's toy to large-scale industrial goods, and the data tells them how the products are used in the real world. This kind of knowledge allows for better design of these products in the future.

Big data will assist in creating new opportunities for growth, and it will also help to create entirely new categories of companies, i.e., those that aggregate industry data and analyze it. A good proportion of these companies will be situated in the center of large flows of information, where the data that come from products, buyers, services, and suppliers can be analyzed. Now is the time for forward-thinking managers and company leaders to build up their company capabilities for big data, and they need to be aggressive about it.

The scale of big data isn't the only important factor; we must also take into account the high frequency and the real-time nature of the data as well. Take "nowcasting" for example. This is the process of estimating metrics, like consumer confidence, straight away – something that at one time could only be done after the fact. This is being used more extensively, thus adding considerable potential to prediction. In a similar way, high frequency allows users to analyze and test theories at a level that has not been possible previously.

Studies of major industries have shown that there are a few ways through which big data can be leveraged:

1. Big data has the potential to unlock some serious value for industries by making all information transparent. There is still a lot of data that has not yet been captured and stored in digital form or that cannot easily be found when searched for. Knowledge workers are spending up to 25% of their time and effort in searching for specific data and then transferring it to another location, sometimes a virtual one. This percentage represents a vast amount of inefficiency.

2. As more and more companies store their transactional data in a digital format, they are able to collect detailed and highly accurate performance information on just about everything – from inventory right down to the amount of sick days being taken – thus giving

themselves the ability to boost performance and root out variabilities. Some leading companies use this ability to collect big data and analyze it to conduct experiments and see how they can make more informed management decisions.

3. Big data allows companies to divide their customers into smaller segments, allowing them to better and more precisely tailor the services and products that they offer.

4. More sophisticated analytics allows for much better decision making. It can also cut down risks and bring to light some valuable insights that might otherwise have never seen the light of day.

5. We can use big data to create the next generation of services and products. For example, manufacturers are already using the data that they get from their embedded sensors to come up with more innovative after-sales service.

Big Data Creates Value

Let's use the U.S. healthcare system as an example here. If they were to make effective and creative use of big data to improve quality and efficiency, they could actually create in excess of $300 billion of value every single year. Around 70% of that would be from a cut in healthcare expenditure of around 8%.

Moving on to the European developed economies; if government administrations used big data in the right way, they could create improvements in operational efficiency of around €100 billion every year. That is just in one area; we haven't looked at what they could achieve if they used advanced analytic tools to boost tax revenue collection and cut down on errors and fraud.

It isn't just organizations or companies that gain benefits from using big data. The consumer can benefit significantly, i.e., those who use services that are enabled by location data can realize consumer surplus of up to $600 billion.

Take smart routing that uses real-time traffic information for example. This is one of the most used of all the applications that employ location data. As more and more people use smartphones and more of those people take advantage of the free map apps, smart routing use is likely to grow significantly. By the year 2020, it is expected that more than 70% of all mobile phones will have GPS capability built in, way more than the 20% recorded in 2010. As such, we can estimate that by 2020, the global value of smart routing has the potential to create savings of around $500 billion in fuel and time. This is the equivalent of cutting 20 billion driving hours or saving a driver around 15 hours a year on the road and savings of around $150 billion in fuel.

The most potential for value from big data is going to come from the combination of data pools. Again, the U.S. healthcare system has four significant data pools – activity and cost; clinical, medical, and pharmaceutical products; R&D; and patient data. Each of these data pools is captured and then managed by a different part of the healthcare system, but if big data were used to its full potential, annual productivity could be increased by around 0.7%. This would require the combination of data from all the different sources, including those from organizations that do not share data at scale. Data sets, such as clinical claims and patient records, would need to be integrated.

Doing this would realize benefits for everyone, from the industry payers to the patients. The patients themselves would have better access to more healthcare information, and they would be able to compare the process of physicians, treatment, and drugs. They could compare effectiveness, allowing them to pick the medication and treatment that suit them the best. To be able to take advantage of these benefits, though, they would have to accept a tradeoff between the benefits and their privacy.

Data security and privacy are two of the biggest hurdles that must be faced if the benefits of big data are to be truly realized. The most pressing challenge is the huge shortage of people who have the skills needed to be able to analyze big data properly. By 2018, the U.S. is facing a shortage of between 140,000 and 190,000 people with the right training in deep analysis and another shortage of around 1.5 million people with the quantitative and managerial skills needed to interpret the analyses correctly so that they can base their decisions on them.

There is also a slew of technological issues that will have to be resolved. Incompatible formats and standards, as well as legacy systems, often stand in the way of data integration and stop sophisticated analytical tools from being used. Ultimately, in order to make full use of the larger digital datasets, a technology stack being assembled from computing and storage right through to the application of visualization and analytical software will be required.

Above all, to take true advantage of big data, access to all data has to be widened. More and more organizations will need to have access to data stored with third parties – for example, with customers or business partners – and then integrate that data with theirs. One of the most important competencies in the future for data-driven companies will be the ability to come up with compelling value propositions for other parties, like suppliers, consumers, and possibly even competitors to a certain extent.

For as long as the true power of big data is understood by governments and companies, the power it has to deliver better productivity, more value for the consumer, and the power it has to fuel the net wave of global economy growth, there should be some incentive for them to take the necessary actions to overcome the barrier that stand in their way. By doing this, they will open up new avenues of competitiveness among industries and individual companies. They will create a much higher level of efficiency in public sectors that will allow for better services. Even when money is short, they will enable more productivity across the board.

Big Data Brings Value to Businesses Worldwide

The value big data has brought and will continue to bring to businesses worldwide is immeasurable.

Here is just a brief summary of ways that it has impacted our world:

- It has created a whole new career field – Data Science
- Big data has revolutionized the way data interpretation is applied
- The healthcare industry has been improved considerably by the application of predictive analytics
- Laser scanning technology has changed the way law enforcement reconstructs crime scenes

- Predictive analytics is changing the roles of caregivers and patients
- Data models can now be built to investigate and solve many business problems
- Predictive analytics is changing the way the real estate industry is conducting business

Big Data = Big Deal

One more thing that big data could do is open the way for new management principles. In the early days of professional management, corporate leaders found that one of the key determining factors of competitive success was a minimum scale of efficiency. In the same way, the competitive benefits of the future are likely to build up with companies that can capture more data that is of higher quality and then use that data at scale with more efficiency.

The following five questions are designed to help company executives determine and recognize just how big data can benefit them and their organizations:

- **What will happen in a "transparent" world, with data available so widely?**

Information is becoming more and more accessible across all sectors, and as such, it has the potential to threaten organizations that rely heavily on data as a competitive asset. Take the real estate industry, for example. They trade on access to transaction data and a secretive knowledge of bids and buyer behaviors. To gain both requires a great deal of expense and a lot of effort. However, recent years have shown that online specialists have begun bypassing the agents and allowing buyers and sellers to exchange their own perspectives on property values and have created a parallel resource for real estate data to be garnered from.

Pricing and cost data have also become more widely accessible across a whole slew of industries. Some companies are now assembling satellite imagery that is readily available; imagine that, when it's processed and then analyzed, such data can provide clues about the physical facilities of their competitors. This gives them ideas about the expansion plans or any constraints that their competitors are coming up against.

One of the biggest challenges is that much of the data that is being amassed is being kept in departmental "silos," such as engineering, R&D, service operations, or manufacturing. These stop the data from being exploited in a timely manner and can also cause other problems. Financial institutions, for example, suffer because they do not share data among the diverse business lines, such as money management, financial markets, or lending. This can stop these companies from coming up with a coherent view of their customers, on an individual basis, or of having an understanding of the links between the financial markets.

Some manufacturing companies are trying their hardest to get into these enclaves. They integrate data from a number of different systems, inviting formerly closed off units to collaborate, and looking for information from external parties, such as customers and suppliers, to help co-create future products. In the automotive industry, global suppliers make hundreds of thousands of different components. Integrating their data better would allow the companies, along with their supply chain partners, to collaborate at the design stage, something that is crucial to the final cost of manufacturing.

- **Would being able to test your decisions change the way a company competes?**

Big data brings in the possibility of a different style of decision-making. Through the use of controlled experiments, organizations will be able to test out hypotheses and then analyze the results. This should give them results that they can use to guide their decisions about operational changes and investments. Effectively, experimentation is going to allow managers to distinguish between correlation and causation, cutting down on the wide variations in outcomes while boosting products and financial performance.

These experiments can take a number of different forms. Some of the top companies online are always testing and running experiments; in some cases, they will set aside a specific part of their web page views in order to test which factors drive sales or higher usage. Companies that sell actual physical products use tests to help them make decisions, but using big data can take these experiments to a whole new level. McDonalds has fitted devices into some of its stores to track customer interaction, the amount of traffic, and patterns in ordering. Using the data gained, they can make decisions on changes to their menus, changes to the design of their restaurants, and training on sales and productivity, among other things.

Where it isn't possible to carry out controlled experiments, a company could use a natural experiment to determine where the variables in performance are. One government sector collected data on a number of different groups of employees who were all working at different sites but doing similar jobs. Just by making that data available, the workers who were lagging behind were pushed to improve performance.

- **If big data were used for real-time customization, what effect would it have on business?**

Companies who face the public, as it were, have been using data to divide and target specific customers for a long time. Big data takes that much further than what always used to be considered as the top of ranged targeting by making it possible to use real-time personalization. In the future, retailers will be able to keep track of individual customers and their behaviors by monitoring internet click streams. They will also be able to make changes to preferences and model the potential behavior in real-time. Doing this will make it easier for them to know when a customer is heading towards making a decision on a purchase; they will be able to "nudge" the decision and take the transaction through to completion by bundling together products and offering benefits and reward programs. This is called real-time targeting, and it will also bring in data from loyalty programs, which, in turn, can help to increase the potential for higher-end purchases by the most valuable customers.

Retailing is probably the most likely of all industries to be driven by data. They have vast amounts of data at their fingertips – from purchases made on the internet, conversations on social network sites, and from location data from smartphone interactions, alongside the birth of new, better analytical tools that can be used to divide customers down into even smaller segments for better targeting.

- **How will big data help management, or even replace it?**

Big data opens up more avenues for the application of algorithms and analysis that are mediated by machines. Some manufacturers use algorithms to analyze data garnered from sensors on the production line. This helps them to regulate their processes, reduce waste, increase output, and avoid human intervention that can be expensive and, in some cases, dangerous. In the more advanced "digital oilfields," sensors are used to monitor the conditions of the wellheads, pipelines, and mechanical systems constantly. That data is then analyzed by computers, and the results are fed to operation centers where the oil flows are adjusted to boost production and reduce downtime in real time. One of the biggest oil companies in the world has managed to increase their oil production by 5% while reducing staffing and operating cost by between 10% and 25%.

Products that range from a simple photocopier to a complex jet engine are now able to generate streams of data that track usage. Manufacturers are able to analyze the data and in a few cases, fix glitches in software, or send out repair representatives automatically. In some cases, the data is being used to preempt failure by scheduling repairs to take place before systems can go down.

The bottom line is that big data can be responsible for huge improvements in performance and better risk management, as well as uncover insights that would most likely never have been found. On top of that, the price of analytics software, communication devices, and sensors are falling fast, and that means more companies will be able to afford to join in.

- **Can big data be used to create a brand new business model?**

Big data is responsible for coming up with brand new company categories that embrace business models driven by data. A large number of these companies are intermediates in a value chain where they can generate valuable "exhaust" data that is produced by transactions. A major transport company saw that while they were going about their transport business, they were gathering large amounts of data on product shipments across the globe. Sensing a top opportunity, they came up with a niche that now sells the data gained to supplement economic and business forecasts.

Another major global company learned a great deal by analyzing their own data. They then decided that they should create a new business to do the same sort of work for other organizations. They now aggregate supply chain and shop floor data for a large number of manufacturers and sell the relevant software tools needed to improve performance. This part of the business is now outperforming their main manufacturing business, and that is all thanks to big data.

Big data has created a whole new support model for existing markets. With all the new data needs of companies, the increased demand for qualified people to support that data is increasing.

As a result, your business may need an outside firm to analyze and interpret the data for you. These are specialized firms that can assimilate large amounts of data, both in structured and unstructured models. These companies will exist solely for the purpose of supporting leading companies in whatever industry. Their employee base would be trained to look for and gather data in systems and processes. These data analysis companies support your business by doing this function for you and charging for their expertise.

They would have the resources and training to assimilate, analyze, and interpret new trends in the data, and they will be required to report any notifications back to you.

A viable alternative would be for existing companies to employ data support departments within their existing infrastructure. This method would be much more cost-effective than using an outside firm but will require specialized skill sets within your company. This could be the next step from your current information technology group. The data support analysts only focus on data in the existing information systems. Their focus would be deciphering data flow by analyzing, interpreting, and actually finding new applications for this data. These new applications and support team would monitor existing data for any fraud, triggers, or issues that it might present.

Big data has created a whole new field of study for colleges and higher institutions of learning to offer in their degree programs. People are being trained in the latest methods of big data gathering, analyzing, and interpreting. This career path will lead them to critical positions in the newly trending data support companies. Big data has not only created a new industry, but a whole new field(s) of study. Along educational lines, big data will change the way teachers are hired. Big data recruiting processes, combined with predictive analytics, can stimulate interactive models between the traits that the most effective teachers should have to maximize the learning experience of students.

Chapter 3: Big Data Development

While the vast majority of data ever collected and analyzed have come in the past few years, the term "Big data" has been present since 2005, and the analysis of data, in general, has quite literally been around for thousands of years. Ever since accounting was introduced back in ancient Mesopotamia to track the increases and decreases of herds and crops, data analysis has always been used in some manner.

The first large-scale statistical data analysis came in the 17th century, when John Graunt published his "Natural and Political Observations Made upon the Bills of Mortality. This book was revolutionary as it used data to provide magnificent insights into the causes of death in 17th century London, and the information inside was meant to help stop future occurrences of the Bubonic plague.

While Graunt's book and his approach to things were a revolution, and statistics as such was now invented, there was no real way of applying it before the invention of computers. This is why data analysis really came into its own in the 20th century with the start of the age of information.

The first ever computing machine that could analyze data was invented in 1887 by Herman Hollerith, who used it to organize census data. The first big data project was organized by the Franklin D. Roosevelt administration in order to keep track of social security contributions for millions of Americans using punch cards and punch card reading machines.

The first real data processing machine came as one of the inventions of World War 2, when British intelligence wanted to decipher German Nazi codes. This machine was dubbed Colossus and could process 5000 characters per second to find patterns in coded messages. This way, the task was reduced from weeks of manual labor to just hours of machine time. This was a real victory of technology and the first massive improvement to statistical analysis.

Electronic storage of information started in 1965. It was again the American government that came up with the idea, and this time, the plan was to store all the tax return claims and fingerprints on magnetic tapes. This project did not get completed because the American public was afraid of such a massive data center as the fear of "Big Brother" was ever present, but the age of electronic data storage was now underway, and no one would be able to stop it.

It was, of course, the invention of the World Wide Web or the Internet that sparked a true revolution in the data storage and analysis area, as Tim Berners-Lee could hardly have a clue as to how immensely he would change the world when he developed his hypertext sharing system. It was the 90s that turned this fledgling technology into the monster it has become today. In 1995, the first supercomputer was made, and it could do as much data analysis in a single second than a human with a calculator could do in 30,000 years. This was the next immense stride in data analysis, as now there was finally a machine out there that could basically replace human labor once and for all.

The term big data first appeared in 2005, when Roger Mougalas mentioned it in referring to any set of data that is so large that it could not possibly be analyzed by using traditional business intelligence tools. That same year, Hadoop was invented by Yahoo to index the entire internet, and today, the tool is used by many companies to crunch through their own sets of big data.

Eric Schmidt famously said in 2010 at the Techonomy Conference in Lake Tahoe that the amount of information that was created by humans between the dawn of civilization and 2003 (about 10 exabytes) was equal to the amount that is created by humanity in just two days in 2010. This piece of information was truly revealing as to how important data had become in our lives.

Since then, big data has only become a bigger part of our everyday lives, as hundreds of new upstarts are attempting to take on big data, and thousands of businesses are using big data in one way or another to optimize their business models. Industries from agriculture to marketing and finance are using the inferences made by analyzing big data, and the world is increasingly facing a shortage of data analysts and scientists. In the age of information, it is no wonder that data has become the most valuable currency, and those who are able to properly use it are those who profit the most.

We can anticipate a major growth in the big data industry over the coming decades. As our world becomes more and more entwined with things like social media, which brings us ever closer to each other and making our lives ever more public, it is no wonder that data collection will become even more relevant. In the foreseeable future, it is certain that companies will keep trying to use big data to infer new ways through which they could sell us products and services, and our governments are bound to keep using it, not only to improve our environment but also get our votes and keep us in check.

Thanks to big data, the future is truly a mystery, as it may go either way. On one hand, big data has the potential to change our world for the better in so many ways, but on the other hand, the way private corporations are using it today may also end up hurting us all. A larger problem is that big data gives more and more power to our governments, and if not used properly, it may lead to the realization of much of the Big Brother fears.

Chapter 4: Weighing the Benefits and Drawbacks of Big Data

It used to be that a recession or economic crisis would sweep the land, and the best anyone could tell you about it was that we have hit a bad patch. Today, thanks to big data, we are able to give extremely precise measurements of various economic, social, and other elements and make statements such as, "This fiscal year, the economy has shrunk by 0.5%." This kind of amazing progress in our ability to measure and analyze data has amazing implications that can be both positive and negative.

The debate on big data and how exactly it might influence the world in future decades has been ongoing ever since big data as a term was introduced. Quite clearly, as we will see in further chapters, there are countless positive ways in which big data could change our everyday lives. Moreover, various economic and social studies experts have also brought to our attention various drawbacks that big data may present us with in the future or has already begun presenting us with, the least of which is the ever-diminishing sense of privacy as more and more public and private organizations gain access to very detailed information on various aspects of our lives.

The politicians in Washington are well aware of this ever-growing concern that big data may have many negative effects on the daily lives of average Joes. Reports from the White House have addressed this issue, claiming that big data analysis needs to find a balance between the social and economic value it presents and the privacy and fairness it may be violating in order to become the driving force of economic and social progress in the coming years.

So let us examine some of the benefits and drawbacks that big data presents in today's world.

The Benefits

We all know that big data can help both the private and public sectors in many ways, but what exactly are these ways? Below are some of the most common ones.

Creating New Streams of Revenue

This one may only be of use to company owners and employers, but a strong economy also means that more people can keep their jobs. This is why it is important for companies to actually make a profit from their work in order to continue employing people.

Analyzing big data can open new frontiers for companies in all sorts of sectors. The big data companies collect information that is valuable not only to them but also to other companies, so trading big data can be another great way to generate even more revenue by selling the data itself or the completed analysis of such data to other companies.

Improving Healthcare

One element of the public sector that really benefits greatly from big data is healthcare. By analyzing immense amounts of information about patients, hospital staff, and the needs and wants of the public in the healthcare sense, healthcare experts are now able to understand better than ever what exactly that the public needs from them.

Even more importantly, merging big data and genetics is one of the ways big data will revolutionize the world. What if every person's health records held their entire genetic map within? What if we could analyze hundreds of thousands of these genetic maps and find the particular genes that are to blame for various illnesses? The application of big data analysis in healthcare is truly limitless, and there is no telling how far we can go with using this kind of technology and data analysis combination.

While most of other industries have gone to great lengths to try and personalize their services, healthcare has been lagging behind, but the big data revolution will undoubtedly change all this and bring us closer than ever to personalized medicine.

Improving Our Environment

How many trash cans do we need per city block? How much street lighting do we require? What time of day can we expect major blocks in traffic? Does a street require an extra lane? These and countless other questions about our cities and environment can be answered by using big data analysis. Thanks to various surveillance systems, we are able to gather any amount of data on what happens in public spaces, such as streets, parks, and highways. The analysis of this type of data can not only save billions of dollars in budgets but also make our lives significantly better in very concrete ways.

The use of big data has already revolutionized many cities around the world. For instance, Oslo, Norway was able to significantly reduce the amount of power they use for street lighting. Meanwhile, Portland, Oregon was able to use big data analysis to significantly reduce their CO2 emissions, while the police department in Memphis, Tennessee reduced the rates of serious crime in the city by 30% using big data.

Quite clearly, big data can help us revolutionize our cities. Imagine if every single aspect of our cities was connected to a central mainframe that could process the big data in real time. Everything from water and electricity consumption, traffic, delivery systems and security of our streets could be improved dramatically through the proper usage of big data. Without a doubt, more and more cities will continue to introduce these methods to improve their systems, and eventually, the entire world will be run by big data.

Making Faster and Better Decisions

No matter what industry we are talking about and regardless of whether the final result is increased revenue, security, or better health for everyone, big data simply allows us to act faster. Analysis of big data allows any sector to make better and more informed decisions, whether it be on how to approach customers or how to improve the healthcare system to get everyone faster access to the treatments they need.

As big data analysis progresses, we become increasingly are to analyze it in real-time or close to real time, which translates into extremely fast results and allows us to make decisions based on the most updated data.

New Services and Products

In the past, companies used to create products based on what they thought people might like to see. Today, thanks to big data analysis, we are able to find out so much more about people's interests that this is no longer the case. For instance, Google Trends is one service that provides companies with the exact data on what it is people are looking for on the World Wide Web through the use of search engines to look up products. By having this kind of information, companies are able to introduce new services and new products that are more personalized to the needs of the customers than ever before.

The Drawbacks

As I previously mentioned, not all is so fine and shiny when it comes to big data. With all the improvements big data is able to bring to both the economy and our personal lives, it is difficult to argue against it, but there have been more than a few complaints, and they have certainly been legitimate. Sometimes, such technology as big data is like a double-edged sword.

Privacy

The use of big data has been mostly criticized by civil rights activists and those who believe that our right to privacy is much more important than the potential advantages that big data could bring. For instance, the collection of big data to compare one user's personal data with a huge set of data about the general public can tell companies a great many things about an individual user. For instance, marketers can use big data to push products toward unsuspecting users by using their very subconscious minds against them. There are many ways in which a marketer can persuade us to buy products that we do not really need by playing with our deep desires and things that we would never normally tell them. Many civil rights advocates say that big data analysis, especially by the private sector, is simply too big of an intrusion into our personal lives.

This argument is really hard to bring down and has considerable merit. For instance, studies have shown that accurate big data analysis can actually tell a company which color their product should have in order to incentivize people to buy it based on a variety of factors. This truly sounds like a magic trick and is surely not one you would want anyone pulling on you, so the argument is that big data collection and analysis by the private sector should simply not be allowed.

The Big Brother

Over the years, the concept of Big Brother has been a very present argument for everyone, from the average, everyday people to conspiracy theorists. For quite some time now, we have known that our governments have been watching us and doing all sorts of things to keep us "in check." Some conspiracy theorists go so far as to say that a small group of extremely powerful men and women are now running the entire world, but even the most moderate of us still understand that governments do collect a huge amount of data that they might not really have the right to collect.

The fear of the Big Brother is something that is very prominent in Western societies, but as time progresses, it seems that this is becoming more and more real. For instance, it is now nearly impossible to walk several blocks of any American city without being filmed by numerous cameras. There is also the topic of the GPS devices that are on our phones and vehicles. Satellite footages are becoming more and more available to governments, and the question we have to ask ourselves is: Am I ever alone?

While this Big Brother style of data collection has led to a lot of good, such as the previously mentioned decrease in crime rates, it is quite natural for people not to want to be monitored all the time, even when they are doing nothing wrong. Recent leaks in American Intelligence have told the public that our phones are being tapped, our social media correspondence is being tracked, and basically, everything we say or do is available to some government agency somewhere. This gives many people who are not doing anything suspicious a feeling of uneasiness.

People are afraid that if the government has this much information, then it holds too much power – even more than the private sector – as everyone is well aware that in this day and age, information is the key to all power.

This is why the regulation of big data analysis for public agencies is paramount. While our system may seem democratic, there is simply plenty of room for public control if the government has so much information, and we should at least be asked what part of the information we want or do not want to share with them.

Crushing Small Businesses

Sure, small businesses can still use big data, but with the amount of power and resources big corporations have, there is no way that a small business can compete with a corporation whose analytics team consists of dozens if not hundreds of data scientists. While the age of corporations has long started, up until a few years ago, small businesses still had ways to compete, and personalizing their service was one of them. Today, with the amount of extensive personal data corporations can get from big data, there is almost no way for a small business to ever offer anything to a customer that a big corporation is not already doing.

Data Sharing and Security

What is even worse is that all of this data is being stored on computers that can be accessed through the internet. While it may be bad that a single corporation has my information, what happens when they decide to share it? What happens when it gets stolen from them? These are all very real concerns.

For starters, it is no secret that companies share either specific user data or big data analysis, whether legal or not. By doing this, our personal data becomes even more exposed, and there may now be many companies out there, which we have never even had any dealings with, that may have our complete user profiles.

Even more disturbingly, servers can get hacked. What happens when an exterior hacker accesses millions of user profiles with pictures, email addresses, and other sensitive information? What about our credit card numbers?

There is simply no way to truly protect the data that is collected for the purposes of big data analysis, which means that this type of collection may need to be limited in the first place.

Wrong Data and Inaccurate Analysis

We have established that big data has a major potential in terms of improving various types of business ventures, but this can only happen when the right data is collected and when it is processed in the right way. However, since neither the data collection systems nor the data analysis systems are anywhere near perfect at the moment, relying too much on big data analysis can actually severely harm businesses.

For starters, the data collection system that is currently in place could be flawed. Often, these systems will not collect a sufficient amount of relevant data, or the data will be skewed one way or another by a host of factors. Even if the analysis of such data is correct, you will actually end up with incorrect final data and potentially make wrong inferences, which could cost the business millions of dollars.

Furthermore, analysis systems are imperfect and still subject to human error. If you put all of your trust in data analysis results, you could end up harming your business, given that such data does not necessarily have 100% accuracy.

In fact, some studies have shown that over 15% of all data intelligence projects at the moment are so wrong that they may put the very existence of their companies in jeopardy. Approximately 30% of other companies will also end up showing a loss, so it is very important to be careful with your big data analysis results and not jump too deep too fast.

Final Look

In conclusion, big data has both good and bad sides, and like any other major social issue, it needs to be observed in a realistic way, and the pros and cons must be weighed. While there are countless ways in which big data can and has already been making our lives better, there are also major concerns.

There is no doubt that big data will be subject to many congressional discussions and other political debates, and it remains to be seen in which ways it will be dealt with. The one thing that is certain, though, is that our lives are becoming less and less our own by the day, and there is no point in denying this.

As we are monitored more and more, it is important to remain aware of the fact that the majority of data collected by the public agencies is being used to improve our lives and decrease budgetary spending. Private sector data collection, on the other hand, is usually voluntary, that is, we provide the companies with our information ourselves. Whether or not they should use it the way they have been is up to the legislature to determine.

Chapter 5: Benefits of Big Data to Small Businesses

One reason small businesses watch from a distance as certain things happen is that often, the cost involved is prohibitive. As for big data, a good number of large-sized companies have embraced it, and that has created the notion that small businesses are yet to qualify for big data. Nothing could be further from reality.

With data analytics, there is no need for complex systems that come with huge monetary demands if the enterprise in question is small. Much that is needed is based on proper organization, human attention to detail, as well as how well people in the organization are equipped in the skills of data analytics. Any organization that can gather ample data that is relevant to its operations and analyze it critically has great chances of seizing business opportunities that they would otherwise have missed.

A small business can, therefore, utilize data analytics to improve the quality of its products or services, modify or fully alter its marketing strategies, and even improve on customer relations. All those areas have a great impact on the overall productivity of the business and, hence, its sustainability and profitability. However, such improvement does not necessarily come with a massive price tag.

Examples of Cost-Effective-Techniques of Data Analytics

Why is cost an issue, anyway? The reason cost is always

factored in before a crucial business decision is made is that cost has a direct impact on profitability. If, for instance, your small business increases sales revenue by 30% at a cost that has risen at 35%, the net profit will not have increased, and the newly introduced techniques will have backfired. Still, you cannot stay put when your competitors are reaping increased profits from improved systems, whether those systems have anything to do with data analytics or not.

If you realize your competitors are using data analytics and, before you know it, they are eating into your market share, it would be suicidal not to consider the use of data analytics, too. At the end of the day, big data brings to the forefront any business opportunities that exist, and those opportunities come with reduced costs and increased revenues. Thus, it is wise for small entrepreneurs to consider using it. After all, small businesses grow by making use of new business opportunities in the same way big businesses do.

This is what small businesses ordinarily rely on for growth:

- Personal intuition
- Commitment and ability to provide high-quality service

Why Small Businesses Do Not Often Give Priority to Big Data

All businesses, large or small, have run their operations traditionally for a long time. However, when developers come up with innovative ideas, they first target large businesses, as those are likely to jump at new ideas simply because they can afford to implement them, cost notwithstanding. The scenario is not much different for big data. Many vendors who market enterprise software solutions put a lot of emphasis on the advantages of economies of scale. Now, how can a small business even think of economies of scale, when it's actually struggling to build capacity?

With such kinds of marketing, small enterprises consider themselves below par, as if big data is not for them. This may have made business sense when the idea of big data has not yet become very popular, especially because a relatively big capital outlay was required upfront in preparing to roll out the application to the thousands of end users involved. This is definitely where economies of scale come in.

However, in today's heightened level of business competition, small businesses need to borrow a page from the world's leading companies, such that if big data is working for huge enterprises, small businesses may try it, too. Luckily, innovators have come up with data solutions that are suitable for small businesses – solutions that equip the small businesses with the appropriate tools to be used by the type of end users that they engage. These solutions are designed to help a small business accomplish its work faster and more efficiently.

Unless small businesses begin to adopt the concept of big data, large competitors are likely to expand their market and run the small entrepreneurs out of town. After all, when big companies use big data, they increase their efficiency, which gives them the drive and ability to embrace a bigger market share at the expense of small companies. As a consequence, it becomes increasingly difficult for small businesses to break even after resigning themselves to the fate of continually declining profits.

Areas Where Big Data Can Help Small Businesses become Cost Effective

- Social media

Here, the small business could analyze the relationship that exists between the use of social media, such Twitter, Pinterest, and Facebook, and users' tendency to consume products or services. With this kind of analysis, you can proceed to capitalize on emerging sales opportunities. By setting up a business presence on these social platforms and targeting potential clients/customers using the site's consumer analytics and ad platforms, you can grow your client base faster than when you use traditional marketing strategies. Implementing special targeting ad strategies can place your ads in front of the clients and customers that are the most interested in your products and services. This is a cost-effective marketing strategy when done correctly. However, some specialized knowledge of each platform is needed to make this effective.

- Launch of an online service

You can take the opportunity to analyze people's tendency to visit your site. Is the visitor traffic high? If so, how high is it? While at it, you could study the users' habits, such as which pages they click next after the landing page. This is the time you analyze the details that seem to capture the attention of these users. At the same time, you can determine the details that turn off the users.

Aren't such details great at helping you identify the best pages to place your promotional content? Do they not also point at the pages and sites where cross-selling opportunities abound? In short, there are cost-effective ways of utilizing big data, and small businesses can benefit just as much as large ones in this regard.

Factors to Consider When Preparing For a Big Data Solution

1) A customized solution

When you are a small business, you do not go acquiring a big data solution that has been mass produced or one that is suitable for a big business. You need one that your internal users will find applicable and helpful. If your big data solution is not tailor-made for your business, then it needs to at least have capabilities that the users can take as options without being encumbered by useless capabilities that have nothing to do with the needs of the business. It also needs to leverage the solutions your business has already put in place, as well as your entire system.

Ultimately, what you want for your small business is a solution that has everything you need integrated and packaged conveniently for use. Beware of vendors who may wish to have you overhaul your system, removing capabilities already in place that you have adopted and even implemented. Remember to justify every cost you bring upon your business.

In case the recommendations for big businesses seem different from those of small ones, it is only because small businesses do not go for a massive integrated solution for the entire organization. Rather, they work with smaller cost centers, such as departments. For example, the marketing department could requisition for a marketing automation system without relying heavily on the views of the IT department. All that the marketing department is called upon to do is lay out its unique requirements, such as its system requirements, spell out its cost justifications, and then conduct appropriate research to identify the solution that suits the department best. At the end of the day, you could have your small business with a varied range of solutions, each of them befitting its relevant department. Each department collects the type of data it deems relevant to its operations, analyzes it, and acts on it to increase the department's efficiency.

2) Ease of deployment

You need to acquire a big data solution that is simple to deploy. You also need to ensure that you are paying for a solution that the users find easy to use. In short, in terms of putting the solution in place, it need not take more than a couple of days or weeks, including testing. Any big data solution that is bound to take you months or even years to put in place in preparation for use is unsuitable for your small business. For starters, time is money, and a complex solution that will cause a department heavy downtime is not worth it. In any case, as you struggle to deploy your new solution, business opportunities might be passing you by.

As you ascertain simplicity, you also need to ensure that the big data solution you are contemplating on acquiring can work with your other applications. A solution will be worth it only if it can work with your existing applications and systems seamlessly. Otherwise, you could be introducing a liability into your business. You need a solution that users will be able to understand without pushing the organization to outsource analysis of the information to highly paid specialists.

By the same token, the big data solution you identify for your small business needs to be one that your staffers can use without the need to undergo expensive and time-consuming training. In fact, you need to look for a solution with self-service capabilities so that even when you change users, new ones can still make use of it without a hitch and without the need to always call on the IT department to help out.

3) The cost involved

Nothing in business makes sense until you know the cost involved. In the case of big data solutions, it is a good idea to buy one that is versatile so that you can increase the use of its capabilities as your business grows. When the solution is new, you may need to use only a few capabilities, and your solution should allow for that. There should also be room for you to pay only for the capabilities you are prepared to use. However, as the business grows, you may need to use more of the big data solution's capabilities.

Thus, not only should the solution be priced reasonably, but it should also come with a licensing strategy that allows the business to progressively increase the capabilities used as its need for data analytics goes up. Business owners always look forward to their small businesses growing fast, so the big data capabilities you are able to bring on board should match the business rate of expansion, as well as growth.

Clearly, it is possible and practical to switch from being intuition-driven, even when yours is a small business, to being analytics driven. All you need to do is identify an IT solution that is suitable for the size and nature of your business. From then on, your business will be in a position to enjoy the benefits that come with big data, which is mainly the chance to identify viable business opportunities.

Chapter 6: Key Training for Big Data Handling

For the continued success of an organization, management needs to focus on the impact new systems are likely to have on overall operations and, ultimately, the bottom line. Taking a country as an example, the reason that some countries choose to have small armies while others choose to have big ones is that some countries consider the existing threat to their security high, while others consider it low. In short, countries do not enroll their youth into the military just to have them draw a salary. That is the same thinking that runs through the minds of management as they seek to establish the people in their organization who require training in big data.

Regardless of whether an organization is big or small, the ability to employ big data is a worthwhile challenge. Even if various departments may work independently when deploying their applications, big data enables departments to relate better as far as sharing of information is concerned, especially because each of them is able to feed the rest with credible information at a relatively faster rate. Still, it is important that training in big data is limited to people who can actually use it productively in the organization. In short, there needs to be some eligibility criteria when it comes to training.

The highest priority needs to go to employees who handle massive data and who may need to use Hadoop. As explained elsewhere in this book, Hadoop is a software library in open-source form that helps in distributed massive data processing. It is the role of the Chief Training Officer (CTO) to determine who in the organization needs training if big data is to benefit the organization. The officer may consult with various department heads because they understand what each employee under them does, and they know the individuals who handle data in heavy traffic. It is the same way a country's Army General liaises with the government to determine the caliber of citizens who qualify for training and recruitment into the army.

However, the CTO is ultimately responsible for enlisting those persons for big data training, being the topmost person as far as the IT infrastructure in the organization is concerned. The CTO needs to remember that whatever recommendations are taken, they will help establish legacy systems to drive the new technology, which will help the employees to become more efficient in their respective roles.

Traditionally, organizations have tended to place a lot of emphasis on data acquisition. Other times, data is just thrown around, especially now that the web has a whole world of data from every imaginable quarter. However, it has become increasingly clear that massive data is only helpful if there is a way of sorting it out so that specific data is classified and utilized where it is most relevant. Unfortunately, this sorting has been left to be done mainly on the basis of an individual's intuition and judgment, without any structured system. This means there is a deficiency in people's ability to utilize data optimally.

Current Level of Competence in Data Management

A report provided by CIO, a popular site that deals with technological issues, indicates that the Computing Technology Industry Association (CompTIA) has conducted a survey that shows how poorly equipped staff in organizations leveraging data are in the aspect of data management. A good number of them are also deficient in data analysis skills. This survey, which was based on information derived from 500 executives from the business and IT sectors, indicates that half of the organizations in the lead, as far as leveraging data goes, have staff member who are not sufficiently skilled in data management and analysis. When it comes to organizations that only leverage data moderately, 71% of them see their staff as poorly equipped or just moderately skilled in the area of data management and analysis.

What these results point to is the need to make a conscious effort to train staff when it comes to big data, so that the availability of data makes business sense to the organization.

Where, Exactly, Is Big Data Training Required?

1. The IT Department

This is a no-brainer. If there are people who are bugged, raveled, and nagged by the rest of the organization when it comes to matters of IT, it would be the members of the IT department. Even when the liaison of data between departments fails to work properly, these are the people who are called upon to help streamline the systems, so that information flow is better coordinated. It is imperative that they are well versed in matters of big data to be able to give appropriate support to members of other departments. In fact, when the IT department is well trained, they save the organization the expenses it would have incurred seeking outside help in supporting the rolled out systems. The role of IT security has become a new player, given all of the new rules and regulations on how data should be treated and stored. This department also needs to be aware of the latest hacking and breaching applications being used and get in front of these before they happen. There are also departments being built around Data Privacy and Compliance within the IT Department that will need to work heavily with the Legal and Compliance departments to make sure that the data isn't compromised, which may cause significant losses for the company.

2. The Department of Product Development

This department is tasked with not only creating new products but also re-engineering existing products. Everyone involved is called upon to have a fresh way of thinking – what can be aptly termed re-thinking innovation. Everyone in the department is deeply involved in every step of the Research and Development process.

For re-engineering, there is a massive responsibility on the part of the staff, and it involves a full understanding of the product as it currently is and as it is envisaged to be after the innovative process of re-engineering. For this understanding to be comprehensive and helpful, massive data is involved, and it needs to be well analyzed, critically evaluated, and properly utilized. That is why the staff in the department of product development needs to be well trained. They should be in a position to capitalize on the advantages provided by big data, particularly when it comes to data integration across existing touch points in the process of product development. Some of those important touch points include product design, software applications, diagnostics, manufacturing, and quality determination, among others.

3. The Department of Finance

Ever heard something like *show me the money?* Nobody wants to say that more than the staff in the department of finance. If you cannot show them the money, they will want to find out independently if you are really worth funding as a department or if you are actually a liability to the organization. It is clear why the staff in this department needs to be well trained in handling big data. They need to be able to say, in monetary terms, whether there is value in deploying certain resources.

In fact, the department of finance is almost as central to the organization as the IT department, and its employees need to be well trained to make use of big data platforms as far as financial modeling is concerned. Only then can a business be sustained, because without conservative spending and optimum project funding, money runs out, and the business goes under. The essence of training personnel in the finance department is to prepare them to make use of big data in their core roles of business planning, auditing, accounting, and overall controlling of the company's finances. Conversely, when the department does these roles well, it ends up creating massive finances for the organization.

Some of the ways members of the finance staff contribute to the success of the organization when they are well trained in big data are by managing to generate reliable cash flow statements, compliance with finance standards, cost modeling, prize realization, and so on. With the influx of data, the finance function has become more complex, especially when the staff wants to gain insight into the future of the organization. As such, training of finance staff in big data is not a luxury, but a necessity.

4. The Human Resource Department (HR)

Though it may not appear obvious, the Department of HR can utilize skills in the application of big data to improve the quality of personnel hired. Competence in big data analysis also comes in handy in assessing the capabilities of existing employees and in determining the capabilities required in future engagements.

As such, members of the HR department need to be well trained so that they, too, can analyze data according to relevance, with a view of engaging in a more strategic manner when it comes to its functions. With this demand, it becomes imperative that staff members are able to use the available tools for data analysis because such are the competencies that will ultimately help in solving issues, which include staff retention, staff-customer relations that affect sales, talent gaps within the organization, quality of candidates to shortlist, and so on. In short, HR is no longer about the number of employees engaged but also about predictive analysis of the matters that have to do with the human resource.

5. The Department of Supply and Logistics

No business can thrive when customers perennially complain about late deliveries, breakages on transit, and other factors that are irritating to customers. It is important that the staff in the department of supply and logistics be well trained to handle big data so that they can utilize it in implementing the department's strategies and in achieving its goals. Once the members of staff in this department are trained in big data, they will be in a position to improve performance and save on cost, primarily by being faster and more agile in their activities. Overall, their training would greatly improve the department's efficiency in service delivery.

What is the importance of that? Well, it is a massive improvement in customer experience, as well as the emergence of a fresh business model. Needless to say, this achievement comes with significant conservation of resources in terms of reduced downtime, reduced overtime, reduced waste, and so on. Ultimately, the smooth supply chain and the great logistics put in place end up promoting your brand name significantly. This is precisely how a business pulls in more customers, thus increasing the business market share without planned and costly marketing.

6. The Department of Operations

In many organizations, this department also encompasses customer support. As such, it is important to train the employees in this department so that customer satisfaction is considered in whatever an employee does in the course of duty. Moreover, it is important that the employees understand the impact of providing great customer support even after they have made a sale. Big data training enlightens the employees very well in this regard. Being able to analyze big data clearly shows the staff that great customer service and support is key to the success of the organization because it improves customer retention and expands the customer base, among others.

7. Department of Marketing

Do you realize how critical numbers are in marketing? Whether it is the number of customers you have brought on board in the last year, the percentage of market share you have captured within a given time, or the sales revenue your product has earned from a certain demographic, this and more keep the marketing department busy. There is massive data in the market – some positive and some potentially damaging to your brand. It is imperative that members of staff in the marketing department are able to amass the data that appears relevant, analyze it, and sieve it to remain with what can be organized and acted upon to the benefit of your brand. Without the skills to handle big data, it is difficult to make heads or tails of the mass of data comprising the traffic through social media, especially in this era when digital marketing is the order of the day.

However, with proper training in big data, the department of marketing can measure with relative accuracy the response your advertisements have on the market, as well as the volume and impact of click-through-rate, impressions, Return on Investment (ROI), and such other factors. While plenty of such data from social media may look irrelevant to the general web visitors, it is as valuable as gold when it comes to the trained eye of a marketer. Moreover, a trained marketer is able to make good use of the large volume of data generated through all stages of customer interaction, social media, as well as during the sales process. For the marketing team, such platforms as Hadoop are very helpful.

Big data training also helps in retrospection, such that the marketing staff can gauge how well the brand has been doing as compared to competing brands, what it is that other similar brands offer, what varying features competing brands have, and such other information that the business can use to improve on its brand. There is even a function in big data's domain through which the marketing team can crawl competitors' websites and do a bit of helpful text mining.

8. Department of Data Integrity, Integration and Data Warehouse

With the increased amount of data there is to monitor, it is crucial that you have a team of specialists that can take in data from your various company systems. This team will need to be current and trained on the various types of data in the systems, as well as the potential risks associated with them. There also needs to be folks who know how to warehouse all this data and make structured sense of it. From customer protection and privacy laws, the teams that work with and interpret the data will need to know the appropriate rules for handling it.

9. Department of Legal and Compliance

With the new legal and compliance rules surrounding data in large organizations, it is necessary for the legal department to stay aware of new privacy and retention policies as they relate to certain pieces of data. The legal and compliance departments should work together to interpret data privacy laws to ensure that the data is protected, stored, and treated appropriately. Businesses that do not monitor and report their data can suffer significant legal issues and should have policies in place to protect themselves against potential lawsuits and breaches.

Chapter 7: Process of Data Analysis

In this chapter, I highlight the steps involved in the process of data analysis. Let us look at the steps one at a time.

What Is Data Analysis?

Before we discuss the steps involved in the process of data analysis, let us first look at what data analysis is. Data analysis is the process by which raw data is collected from various sources and is then converted into meaningful information, which can be used by various stakeholders for making better decisions.

To put it in the words of John Tukey, a famous Statistician, data analysis is defined as follows:

"Procedures for analyzing data, techniques for interpreting the results of such procedures, ways of planning the gathering of data to make its analysis easier, more precise or more accurate, and all the machinery and results of (mathematical) statistics which apply to analyzing data."

Steps Involved In Data Analysis

Even though the data requirements may not be the same for every company, most of the below steps are common for all companies.

Step 1: Decide on the Objectives

The first step in the data analysis process is the setting of objectives. It is important that you set clear, measurable, and concise objectives. These objectives can be in the form of questions. For instance, your company's products are finding it difficult to get off the shelves because of a competitor's products. The questions that you might ask are, "Is my product overpriced?" "What is unique about the competitor's product?" "What is the target audience for the competitor's product?" "Is my process or technology redundant?"

Why is asking these questions upfront important? This is because your data collection depends on the kind of questions you ask. For instance, if your question is, "What is unique about the competitor's product?" you will have to collect feedback from the consumers about what they like about the product, as well as conduct an analysis on the specifications of the product. On the other hand, if your question is, "Is my process or technology redundant?" you will have to perform an audit of the existing processes and technologies used at your establishment, as well as conduct a survey about the technology used by others in the same industry. Notably, the nature of data collected differs vastly on the basis of the kind of questions you ask. Given that data analysis is a tedious process, it is necessary that you do not waste the time of your data science team in collecting useless data. Ask your questions right!

Step 2: Set Measurement Priorities

Now that you have decided on your objectives, you need to establish measurement priorities next. This is done through the following two stages:

Decide on what to measure

This is when you have to decide what kind of data you need to answer your question. For example, if your question pertains to reducing the number of jobs without compromising the quality of your product or service, then the data that you need in hand right now are as follows:

- The number of staff members employed by the company
- The cost of employing the present number of staff members
- The percentage of time and efforts spent by the current staff members on existing processes

Once you have the above data, you will have to ask other questions ancillary to the primary question, such as, "Are my staff not being utilized to their fullest potential?" "Is there any process that can be altered to improve the productivity of the staff?" "Will the company be in a position to meet increased demands in the future despite the downsizing of manpower?"

These ancillary questions are as important as the primary objective question. The data collected in connection with these ancillary questions will help you make better decisions.

Decide on how to measure

It is highly important that you decide on the parameters that will be used to measure your data before you begin collecting it. This is because how you measure your data plays an important role in analyzing the collected data in the later stages. Some of the questions you need to ask yourself at this stage are as follows:

- What is the time frame within which I should complete the analysis?
- What is the unit of measure? For instance, if your product has international markets and you are required to determine the pricing of your product, you need to arrive at the base price using a certain currency and then extrapolate it accordingly. In this case, choosing that base currency is the solution.
- What are the factors that you need to include? This could again depend on the question you have asked in stage 1. In the case of the staff downsizing question, you need to decide on what factors you need to take into consideration with respect to the cost of employment. You need to decide on whether you will be taking the gross salary package into consideration or the net annual salary drawn by the employee.

Step 3: Collection of Data

The next step in the data analysis process is the collection of data. Now that you have already set your priorities and measurement parameters, it will be easier for you to collect data in a phased manner. Here are a few pointers that you need to bear in mind before you collect data:

- We already saw the different sources of data in the previous chapter. Before you collect data, take stock of the data available. For example, in the case of the downsizing of the staff question, you can just look at the payroll to know the number of employees available. This could save you the time needed for collecting this particular data again. Similarly, you need to collate all available information.

- If you intend to collect information from external sources in the form of a questionnaire, then spend a good amount of time in deciding on the questions that you want to ask. Only when you are satisfied with the questionnaire and believe that it serves your primary objective should you circulate it. If you keep circulating different questionnaires, then you will have heterogeneous data in hand, which will not be possible to compare.

- Ensure that you have proper logs when you enter the data collected. This could help you analyze the trends in the market. For instance, let us assume you are conducting a survey regarding your product over a period of two months. You will note that the shopping habits of people change drastically during holiday seasons, more than any other period of the year. When you do not include the date and time in your data log, you will end up with superfluous figures, and this will affect your decisions significantly.

- Check the budget allocated for the purpose of data collection. Based on the available budget, you will be able to identify the methods of data collection that are cost effective. For instance, if you have a tight budget, and you still have to do a survey to gauge the preferences of your customers, you can opt for free online survey tools as opposed to the printed questionnaires that are included in the packages. Similarly, you can make the best use of social networking sites to conduct mini surveys and collect the data required. On the other hand, if you have an adequate budget, you can go for printed and attractive questionnaires that can be circulated along with the package or that can be distributed at retail outlets. You can set up drop boxes at nearby cafes and malls for the customers to drop these filled out questionnaires. You can also organize contests to collect data while marketing your product in one go.

Step 4: Data Cleaning

Now, the data you have collected will not necessarily be readily usable. This is why data cleaning is crucial in this process ion terms of ensuring that meaningless data does not find its way into the analysis stage. For example, when you correct the spelling mistakes in the collected questionnaires and feed them into your system, it is nothing but data cleaning. When you have junk data in the system, it will affect the quality of your decision. For instance, let us assume 50 out of 100 people responded to your questionnaires. However, you get 10 incomplete questionnaire forms. You cannot count these ten forms for the purpose of analysis. In reality, you have gotten only a 40% response to your questionnaire, not 50%. These numbers make a big difference for the management that will be making decisions.

Similarly, if you are conducting a region-wide survey, you need to be extra careful at this stage because most people have a tendency to not disclose their correct addresses in the questionnaires. Hence, unless you have a fair idea about the population of each region, you will never be able to catch these slip-ups. Why is it important to catch these mistakes? Let's assume that your survey results show that the majority (say 70%) of your customer base is from X region. In reality, the population of the region is not even close to 30% of your customer base. Now, let us assume that you decide to make a decision based solely on your survey results. You decide to launch an exclusive marketing drive in this region. Of course, the marketing drive will not improve your sales because even if all the citizens in the region buy your product, they comprise only 30% of your customer base, not 70% as you imagined. Hence, these little numbers play an important role when it comes to big and expensive decisions.

As you can see, improving the quality of data is highly important for making better decisions. As this involves a lot of time, you should automate this process. For instance, to detect fake addresses, you can get the computer to detect entries that have incorrect or incomplete zip codes. This could be easily done if you are using an Excel sheet to store your data. Alternatively, if you have customized software for feeding and storing data, you can get in touch with the software designer to put certain algorithms in place to take care of such data.

Step 5: Analysis of Data

Now that you have collected the requisite data, it is time to process it. You may resort to different techniques to analyze your data. Some of the techniques are as follows:

- Exploratory data analysis

  This is a method by which data sets are analyzed with a view to summarize their distinct characteristics. This method was developed by John W. Tukey. According to him, too much importance was being placed on statistical hypothesis testing, which is nothing but confirmatory data analysis. He felt the need to use data for the purpose of testing hypotheses. The key objectives of exploratory data analysis are as follows:

  (i) Suggestion of hypotheses in connection with the causes of the phenomena under question

  (ii) Assessing the assumptions on which the statistical inference will be based

  (iii) Supporting the selection of appropriate statistical techniques and tools

  (iv) Providing a basis for further data collection through such modes as surveys or experiments

Several techniques prescribed by exploratory data analysis have been extensively used in the fields of data mining and data analysis. These techniques also form part of certain curricula to induce statistical thinking in students. As you perform exploratory data analysis, you will be required to clean up more data. In some cases, you will be required to collect more data to complete the analysis to ensure that the analysis is backed by meaningful and complete data.

- Descriptive statistics

This is another method of data analysis. Through this method, data is analyzed to identify and describe the main features or characteristics of the data collected. This is different from inferential statistics, wherein the data collected is analyzed to learn more about the sample. These findings are then extrapolated to the general population based on the sample. Meanwhile, descriptive statistics only aims to summarize and describe the data collected. These observations about the collected data can either be quantitative or visual. These summaries could just be the beginning of your data analysis process. These could form the basis on which further analysis is conducted to process the data. To understand this better, let us look at an example. The shooting percentage in the game of basketball is nothing but a descriptive statistic. This shooting percentage indicates the performance of the team. It is calculated by dividing the number of shots made by the number of shots taken. For instance, if a basketball player's shooting percentage is 50%, it means that he makes one shot in every two. Other tools used under descriptive statistics include mean, median, mode, range, variance, standard deviation, etc.

- Data visualization

As the name suggests, data visualization is nothing but the representation of data in a visual form. This can be done with the help of such tools as plots, informational graphics, statistical graphics, charts, and tables. The objective of data visualization is to communicate the data in an effective fashion. When you are able to represent data effectively in a visual form, it helps in the analysis of data and in reasoning about data and evidence. Even complex data can be understood and analyzed by people when put in visual form. These visual representations also facilitate easy comparison. For instance, if you are vested with the job of reviewing the performance of your product and that of your competitor, you will be able to do so easily if all the related data are represented in visual form. All your data team needs to do is use the data pertaining to the parameters, such as price, number of units sold, specifications, etc., and then put it in pictorial form. This way, you will be able to assess the raw data easily. You will also be able to establish a correlation between the different parameters and make decisions accordingly. For instance, if you notice that your price is higher than your competitor's, and your sales are lower than your competitor's, then you know where the problem lies. The decreased sales can be attributed to the increase in price. This can be easily addressed by reworking your prices.

Apart from these three major methods, you can also use software available in the market. The prominent software currently available for the purpose of data analysis includes Minitab, Stata, and Visio. Let us not forget the multipurpose Excel.

Step 6: Interpreting the Results

Once you have analyzed your data, it is time to interpret your results. Here are a few questions that you need to ask:

- Does the analyzed data answer your key question? If yes, how so?
- If there were any objections to begin with, did your data help you defend them? If yes, how so?
- Do you think there are any limitations to your results? Are there any angles that you haven't considered while setting priorities?
- Do you have trained people to interpret data properly?

If the analyzed data satisfies all the above questions, then your analyzed data is final. This information can now be used for the purpose of decision-making.

Importance of Interpreting Data Effectively and Accurately

The importance of accurately interpreting the data cannot be emphasized enough. Your company, website, etc., must have experienced professionals who know how to take organic data and interpret the results properly. For example, let's say your company finds it necessary to analyze data from two of the most popular social media platforms – Facebook and Twitter.

Your company cannot depend on an untrained professional to respond effectively to your "likes" or "tweets" on a minute-by-minute basis. Most companies today employ a Social Media Manager to manage their social platforms. These individuals are trained to know the ins and outs of each social platform and effectively respond to your customers in a way that represents your brand.

At the core of every successful business is the accurate interpretation of vital data. It's necessary to hire professionals who have the training necessary to take the unstructured and otherwise random data and structure it in an understandable manner. This will change the dynamics of how your company operates and what decisions need to be made based on the data.

Trained professionals can take all of your customers' "likes" on your corporate Facebook page and trace the consumers' behavior regarding the use of your product. Follow the decision-making process of these consumers. The consumers like your product, then what? Do they read the product description? Do they reap the benefits of using your product? Is your product reasonably priced in comparison to your competitor's prices? What makes your company's product better than the rest of the competition?

Trained data analysts will be able to trace these questions and analyze the pattern that your consumers will take. They will follow the trace that the consumers make. They can analyze the data from the original "like" from your consumer all the way to the purchase on your website.

The right people who have the training to follow and analyze this process can help your company generate increased product sales by taking this information and disseminating it to the appropriate team members throughout the company. Having properly interpreted and meaningful data may be the difference between your company expanding its influence or shutting down because of misinterpreted data.

An example of how you can analyze tweets is by interpreting

historical tweets and distinguishing the substantial "tweet" from the casual "tweet."

Data interpreters are able to analyze historical data from previous company "tweets" and the influence of such "tweets" on consumer buying habits. These experts can translate which "tweets" are substantial and which "tweets" are just social. From the initial root message texted on Twitter, the analyst is able to trace the impact on the consumer's initial mindset as to whether he or she will help achieve the company's core goal of buying the product. Which text is more convincing than others? Why is it more successful? Do images with the "tweets" tend to convince your consumer base to buy your product? Which "tweets" work best with what regions in the world? Which "tweets" work best with what age group?

These are important questions that can be answered by data, and they show why it is important to have analysts review and identify what marketing strategies are working the best on each platform. Analysts can interpret large amounts of data with the use of visual graphs showing numerical data statistics. These can be given to the appropriate departments so that they can make decisions to improve the overall sales experience for your customers.

Chapter 8: Descriptive Analysis

What is Descriptive Analysis?

Descriptive analysis is the oldest and most commonly used type of analysis by businesses. In business, this type of analysis is often referred to as business intelligence, as it provides the knowledge needed to make future predictions, similar to what intelligence agencies do for governments. This category of analysis includes analyzing data from the past through data aggregation and data mining techniques to determine what has happened so far, which can then be used to determine what is likely to happen in the future.

Descriptive analysis quite literally describes past events, as the name suggests. By using various data mining techniques and processing this data, we can turn such data into facts and numbers that are understandable to humans, thereby allowing us to use this data for planning our future endeavors.

Descriptive analytics allows us to learn from the past events, whether they occurred a day or a year ago, as well as to use this data to anticipate how they might influence future behavior. For instance, if we are aware of the average number of product sales we made per month in the past three years and are able to see trends like rising or falling numbers, we can anticipate how these trends will influence future sales, or we can see that the numbers are going down. This means we will need to change something in order to get the sales back up, whether it means re-branding, expanding our team, or introducing new products.

Most of the statistics businesses use in their everyday

operations fall into the category of descriptive analysis. What statisticians will do is to collect descriptive data from the past and then convert them into a language understandable to the management and employees. Using descriptive analysis allows businesses to see such things as how much they are spending on average on various expenses, how much of the product sales percentage falls to expenses, and how much is clear profit. All of these allow us to cut corners and make more profits in the end, which is the name of the game in business.

How Can We Use Descriptive Analysis?

Descriptive statisticians usually turn data into understandable output, such as reports with charts that show what kind of trends a company has seen in the past in a graphical and simple way, thus allowing this company to anticipate the future. Other data include those regarding a particular market, the overall international market, or consumer spending power, etc.

A good example of descriptive analysis can be a table of average salaries in the USA in a given year. A table like this can be used by various businesses for many purposes. This particular example of statistical analysis allows for deep insight into the American society and individuals' spending power and has a vast array of possible implications. For instance, from such a table, we could see that dentists earn three times more money than police officers, and such data could possibly be useful in a political campaign or in determining your target audience for a given product. If a business is a fledgling, for example, they could make a vast number of decisions about their business plan on the basis of this table.

Values in Descriptive Analysis

There are two main ways of describing data, and these are measures of central tendency and measures of variability or dispersion. When we are talking about measuring a central tendency, we basically mean measuring data and finding the mean value or average from a given data set. This mean is determined by summing up all the data and dividing it by the number of data units, getting an average value that can be used in various ways. Another unit used in measuring the central tendency – which is perhaps even more useful – is the median. Unlike the mean, the median takes into consideration only the middle value of a given data set. For instance, in a string of nine numbers, the fifth number is considered the median. If we arrange all our numbers from lowest to highest, the median will often be a more reliable value than the mean because there could be outliers at either end of the spectrum, which bend the mean into a wrong number. The outliers are extremely small or big numbers that will naturally make the mean unrealistic, and the median will be more useful in cases where there are outliers.

Measuring the dispersion or variability allows us to see how spread out the data is from a central value or the mean. The values used to measure the dispersion are range, variance, and standard deviation. The range is the simplest method of dispersion. The range is calculated by subtracting the smallest number from the highest. This value is also very sensitive to outliers, as you could have an extremely small or high number at the ends of your data spectrum. Variance is the measure of deviation that tells us the average distance of a data set from the mean. Variance is typically used to calculate the standard deviation, and by itself, it would serve little purpose. Variance is calculated by calculating the mean, then subtracting the mean from each data value, squaring each of these values to get all positive values, and then finding the sum of these squares. Once we have this number, we will divide it by the total number of data points in the set, and we will have our calculated variance.

Standard deviation is the most popular method of dispersion as it provides the average distance of the data set from the mean. Both the variance and standard deviation will be high in instances where the data is highly spread out. You will find the standard deviation by calculating the variance and then finding its square root. Standard deviation will be a number in the same unit as the original data, which makes it easier to interpret than the variance.

All of these values used to calculate the central tendency and the dispersion of data can be employed to make various inferences, which can help with future predictions made by predictive analytics.

Inferential Statistics

Inferential statistics is the part of analysis that allows us to make inferences based on the data collected from descriptive analysis. These inferences can be applied to the general population or any general group that is larger than our study group.

For instance, if we conducted a study that calculated the levels of stress in a high-pressure situation among teenagers, we could use the data we collect from this study to anticipate general levels of stress among other teenagers in similar situations. Further inferences could be made, such as possible levels of stress in older or younger populations, by adding other data from other studies, and while these could be faulty, they could still potentially be used with some degree of credibility.

Chapter 9: Predictive Analytics

We have seen now how data and data analysis are crucial for the effective functioning of a business. Let us look at another wing of data mining that plays a vital role in the growth of a business. In this chapter, I will be taking you through the various aspects of predictive analytics and help you appreciate the role it plays in facilitating the effective functioning of a business.

What is Predictive Analytics?

In simple terms, predictive analytics is nothing but the art of obtaining information from collected data and utilizing it for predicting behavior patterns and trends. With the help of predictive analytics, you can predict unknown factors, not just in the future but also in the present and past. For example, predictive analytics can be used to identify the suspects in a crime that has already been committed. It can also be used to detect fraud as it is being committed.

What are the Types of Predictive Analytics?

Predictive analytics can be referred to as predictive modeling. In simpler terms, it is the act of pairing data with predictive models and arriving at a conclusion. Let us look at the three models of predictive analytics.

Predictive Models

Predictive models are nothing but models of the relationship between the specific performance of a certain element in a sample and a few known attributes of the sample. This model aims at assessing the likelihood that a similar element from a different sample might exhibit the same performance. It is widely used in marketing. In marketing, predictive models are used to identify subtle patterns, which are then used to identify the customers' preference. These models are capable of performing calculations as and when a certain transaction is happening, i.e., live transactions. For instance, they are capable of evaluating the opportunity or risk associated with a certain transaction for a given customer, thereby helping the customer decide if he or she wants to enter into the transaction. Given the advancements in the speed of computing, individual agent modeling systems have been designed to simulate human reactions or behavior for certain scenarios.

Now, let us look at some more aspects of these models in detail. The terminology "training sample" refers to the sample units/elements that are available, the attributes and performances of which are known. The units/elements that are present in other samples, the attributes of which are known but performances are unknown, are referred to as "out of training sample." There is no chronological relation between the training sample and the out of training sample. For example, the blood splatters in a simulated environment are training samples. On the other hand, the blood splatter from an actual crime scene is the out of training sample. Predictive models can help identify the probable suspects and modus operandi related to the murder based on these samples. As mentioned earlier, these samples are not required to be from the same time frame. Either of the samples could be from a different time frame.

Predictive Models in Relation to Crime Scenes

3D technology has brought big data to crime scene investigations to help police departments and criminal investigators reconstruct crime scenes without violating the integrity of the evidence. There are two types of laser scanners used by crime scene experts:

- Time-of–flight laser scanner

The scanner shoots out a beam of light that bounces off the targeted object. Different data points are measured as the light returns to the sensor. It's capable of measuring 50,000 points per second.

- Phase shift 3D laser scanners

These scanners are much more expensive but also much more effective. They measure 976,000 data points per second. These scanners use infrared laser technology.

These data laser scanners make crime scene reconstruction much easier. Needless to say, the process takes a lot less time than traditional crime scene reconstruction takes. The advantage of 3D technology is that investigators can re-visit the crime scene anywhere. Investigators can now do this while they are at home, in their offices, or out in the field. This makes their job more mobile, and they can visit the crime scenes virtually anywhere. They no longer have to depend on notes or their memories to recall the details of the crime scene. Also, they visit the crime scene once, and that's it. They have all the data images recorded on their scanners.

Investigators are able to re-visit crime scenes by viewing the images on computers or iPads. The distance between objects will be reviewed (like weapons.) The beauty of this technology is that crime experts do not have to second guess the information gathered from crime scenes. The original crime scene is constructed right there on the scanner images. It's as if the crime was committed right there on the scanned images. The images tell the story about who the perpetrators were and how they carried out the crime.

Investigators can look at the crime scenes long after they are released. Nothing in the crime scene will be disturbed or compromised. Any evidence that is compromised is inadmissible in court and cannot be used. All evidence must be left in the original state and should not be tampered with. This is no problem when the evidence is recorded in the data scanners.

Law enforcement engineers are able to reconstruct the whole crime scene in a courtroom. Forensic evidence is untouched and left intact, thus guaranteeing a higher rate of convictions.

Forensic Mapping in Crime Scene Reconstruction

The purpose of 3D forensic mapping is to reconstruct every detail of the crime scene holistically. This is a very pure way of reconstructing crime scene evidence. None of the evidence is touched or accidentally thrown away. Investigators do not have to walk in or around the evidence, thus avoiding the possibility of anything being accidentally dropped or kicked.

3D data mapping allows for greater understanding and insight into the motive and the method of the crime committed. This helps law officials present a convincing understanding in court so that evidence will convince the jury beyond a reasonable doubt that the crime was or wasn't committed.

3D forensic mapping is also invaluable in training new investigators in the process of criminal reconstruction.

Descriptive Models

Descriptive models are used to ascertain relationships in the data collected. This is similar to how a company would classify and group its customers or its products into certain categories. Predictive models focus on predicting only a single customer behavior, as in the case of the computation of the credit risk. Meanwhile, descriptive models focus on identifying different relationships between products or customers. Descriptive models not only seek to rank customers based on their attributes or based on the actions taken by them, as in the case of predictive models. Descriptive models tend to categorize customers based on their product preferences. These models can be used to build more models, which can be used to make more predictions.

Decision Models

The decision model is nothing but a system that contains at least one action axiom. An action axiom is nothing but the action that follows the satisfaction of a certain condition. A model action axiom is as follows:

If <a certain fact> is true, then do <this certain action>.

In simpler terms, the action axiom is used to test a certain condition. The fulfillment of a certain condition necessitates the completion of a certain action.

The decision model also describes the relationship between all the elements that form part of a decision, such as the decision, the known data, and also the forecast results associated with the decision. This model aims to predict the results of those decisions, which have many variables involved in it. Decision models are also used to achieve optimization and maximize certain outcomes while minimizing certain other outcomes. These models can be used to come up with a set of rules for a business. Such rules will be capable of producing the expected and desired action for every customer who purchases the service of the business.

Chapter 10: Predictive Analysis Techniques

In this chapter, let us look at the different techniques used for the purpose of conducting predictive analytics. The two major categories into which these techniques can be grouped are machine learning techniques and regression techniques. Let us look at these techniques in detail below.

Regression Techniques

These techniques form the foundation of predictive analytics. They aim to establish a mathematical equation, which will in turn serves as a model for representing the interactions among the different variables in question. Based on the circumstances, different models can be applied for performing predictive analysis. Let us look at some of them in detail now below.

Linear Regression Model

This model assesses the relationship between the dependent variable in a given situation and the set of independent variables associated with it. This is usually expressed in the form of an equation. The dependent variable is expressed as a linear function of different parameters. These parameters can be adjusted in such a way that it leads to the optimization of measure of fit. Model fitting is required to minimize the size of the residual. Model fitting must be given due importance to ensure that each variable is distributed randomly in connection with the model predictions.

The objective of this model is the selection of parameters with the aim of minimizing the sum of the squared residuals. This is known as the ordinary least squares estimation. Once the model is estimated, the statistical significance of the different coefficients used in the model must be checked. This is where the t-statistic comes into play, whereby you test whether the coefficient is different from zero. The ability of the model to predict the dependent variable depending on the value of the other independent variables involved can be tested by using the R^2 statistic.

Discrete Choice Models

Linear regression models can be used in cases in which the dependent variable has an unbounded range and is continuous. However, there are certain cases in which the dependent variable is not continuous. In such cases, the dependent variable is discrete. Given that the assumptions related to the linear regression model do not hold good completely in the case of discrete variables, you will have to go for another model to conduct predictive analytics.

Logistic Regression

This model is used in cases in which the dependent variable is categorical. A categorical variable is one that has a fixed number of values. For instance, if a variable can take two values at a time, it is called a binary variable. Categorical variables that have more than two values are referred to as polytomous variables. One example is the blood type of a person.

Logistic regression is used to determine and measure the relationship between the categorical variable in the equation and the other independent variables associated with the model. This is done by utilizing the logistic function to estimate the probabilities. It is similar to the linear regression model. However, it has different assumptions associated with it. There are two major differences between the two models. They are as follows:

- The linear regression model uses a Gaussian distribution as the conditional distribution, whereas the logistic regression uses a Bernoulli distribution.
- The predicted values arrived at in the logistic regression model are probabilities and are restricted to 0 and 1. This is because the logistic regression model is capable of predicting the probability of certain outcomes.

Probit Regression

Probit models are used in place of logistic regression for coming up with models for categorical variables. This is used in the cases of binary variables, i.e., categorical variables, which can take only two values. This model is popular in economics. This method is used in economics to predict those models that use variables that are not only continuous but also binary in nature. Two important reasons why the probit regression method is often preferred over logistic regression method are as follows:

1. The underlying distribution in probit regression is normal.

2. If the actual event is a binary proportion and not a binary outcome, then the probit method is more accurate.

Machine Learning Techniques

Machine learning is a field of artificial intelligence initially used for the purpose of developing techniques that will facilitate computers to learn. Given that machine learning consists of an array of statistical methods for classification and regression, it is now being employed in different fields, such as credit card fraud detection, medical diagnostics, speech recognition, face recognition, and stock market analysis. In certain applications, these techniques can be used to predict the dependent variable, without taking into consideration the independent variables or the underlying relationships between these variables. In other words, machine-learning techniques can be readily employed in cases in which the underlying relationships between the dependent and independent variables are complex or unknown. Some of the techniques used in this method are as follows:

Neural Networks

Neural networks are sophisticated and non-linear modeling techniques capable of modeling complex functions. These techniques are widely used in the fields of cognitive psychology, neuroscience, finance, physics, engineering, and medicine.

Neural networks are generally employed when you are not aware of the exact relationship that exists between the inputs and output. However, these networks learn the underlying relationship through training. There are three kinds of training that form part of neural networks, namely, supervised training, unsupervised training, and reinforcement learning. Of the three, supervised training is the most commonly used.

A few examples of training techniques used in neural networks are as follows:

- Quick propagation
- Conjugate gradient descent
- Backpropagation
- Projection operator
- Delta bar delta

Multilayer Perceptron

The multilayer perceptron is made of an input layer and output layer. Over and above these, there are one or more hidden layers made up of sigmoid nodes or nonlinearly activating nodes. The weight vector plays an important role in adjusting the weights of the network. The backpropagation technique is employed to minimize the squared error, which usually arises between the network output values and the expected values for the output.

Radial Basis Functions

A radial basis function is a function that has an inbuilt

distance criterion in connection with a center. These functions are typically used for the interpolation and smoothing of data. These functions have also been used as part of neural networks in place of sigmoid functions. In such cases, the network has three layers, namely, the input layer, the hidden layer with the radial basis functions, and the output layer.

Support Vector Machines

Support vector machines are employed for the purpose of detecting and exploiting complex patterns in data. This is achieved by clustering, classifying, and ranking the data. These learning machines can be used for the purpose of performing regression estimations and binary classifications. There are several types of support vector machines, including polynomial, linear, and sigmoid, to name a few.

Naïve Bayes

This is based on the Bayes conditional probability rule. This is an easy technique that uses constructed classifiers. In other words, this is used for the purpose of classifying various tasks. This technique is based on the assumption that the predictors are statistically independent. Such independence makes it a great tool for classification. Moreover, this technique is also easier to interpret. In cases in which the number of predictors is very high, it is wise to use this method.

K-Nearest Neighbors

The nearest neighbor algorithm forms part of the pattern recognition statistical methods. Under this method, there are no underlying assumptions associated with the distribution from which the sample is drawn. It consists of a training set and has both positive and negative values. When a new sample is drawn, it is classified based on its distance from the nearest neighboring training set. The sign associated with that point also plays an important role in classifying the new sample. While using the k nearest neighbor classifier, all the k-nearest points are considered. The sign of majority of these points is then used to classify the new sample.

The performance of this algorithm is determined by the following factors:

1. The distance measure used to locate the nearest neighbors.
2. The decision rule used to derive a classification from the k-nearest neighbors.
3. The number of neighbors being considered in classifying the new sample.

Geospatial Predictive Modeling

The underlying principle of this technique is the assumption that the occurrences of the events being modeled are limited in terms of distribution. In other words, the occurrences of events are neither random nor uniform in distribution. Instead, other spatial environment factors, such as socio-cultural, infrastructure, topographic, etc., involved. These factors are capable of constraining and influencing the locations of the occurrences of these events. This is a process for assessing events using a geographic filter to come up with statements of the likelihood of occurrence of a certain event.

There are two kinds of geospatial predictive models. They are as follows:

1. Deductive method
2. Inductive method

Let us look at these methods in detail now.

1. Deductive method

   This method is based on a subject matter expert or qualitative data. Such data is then used for the purpose of describing the relationship that exists between the occurrence of an event and the factors associated with the environment. In other words, this method primarily relies on subjective information. The limitation of this model lies in the fact that the number of factors being keyed in is completely dependent on the modeler.

2. Inductive method

   This method is based on the spatial relationship (which is empirically calculated) that exists between the occurrences of events and the factors that are associated with the environment. Every occurrence of an event is first plotted in the geographic space.

Following the plotting, a quantitative relationship is then defined between the occurrence and the factors associated with the environment. This method is highly effective because you can develop software to discover empirically. This is crucial when you have hundreds of factors involved, and there are both known and unknown correlations existing between events and factors. The quantitative relationship values, as defined previously, are then processed with the help of a statistical function. These values are processed to determine spatial patterns capable of defining not only areas that are highly probable for the occurrence of an event but also those that which are less probable locations for the occurrence of an event.

Hitachi's Predictive Analytic System

Would you like to take a look into the future? This is what Hitachi's Predictive Analytic System (PCA) will achieve if it's successful in its unveiling. This technology will revolutionize the way police departments handle crimes. The present method is labeled the "break/fix model." When police arrive on the scene after a crime has been committed, a death, rape, domestic violence, or a burglary has taken place. Then, the police would have to administer damage control. Sometimes, the police can get to the scene before the crime is committed or while the crime is being committed. In this case, they can stop the crime.

What if they could arrive on the scene even before the crime is committed? How many lives would be saved? People could be spared bodily pain from being abused or beaten up. Costly damages from property loss could be reduced or avoided. There are so many benefits that would be derived if the PCA is successful in its intended purpose.

The PCA would be able to predict crimes before they happen. This would enable police departments to arrive on the scene before the crime is committed. Instead of dealing with the quirks of traditional crime predicting methods where investigators use their own experiences (the product of their personal variables) to project the committing of crimes, the PCA will eliminate countless variables to accurately analyze a multitude of factors that will affect crime. Here are the components that the PCA analyzes to predict that a crime will be committed:

- Weather Patterns
- Public Transit Movements
- Social Media Actions
- Gun Shot Sensors
- Many other Factors

The above data sets are fed into the system and over a two-week period, the PCA will determine whether there is a correlation between the said datasets. Hitachi will allow different law enforcement agencies to test their systems in the real-world working environments. These "experiments" will be held in different agency locations in unknown cities. The agencies will participate in what is known as a "double-blind trial." The predictive system will be running in the background, but criminal experts will not be able to see the predictive results as they happen.

The summation of the testing period will be when Hitachi will compare and analyze the results of the predictions in relation to the actual daily police activities over the same period.

Benefits of PCA

The benefits of the success of PCA would be countless:

- Countless lives would be saved
- Law enforcement agencies could still work efficiently in the light of continual manpower cuts.
- Officers can be placed strategically in locations to stop potential crimes
- Officers would exponentially improve their crime-fighting time exponentially
- Law enforcement training costs would dramatically drop
- Law enforcement experts would be in less life-threatening circumstances
- Homicides, rapes, and domestic violence cases would drop
- Stolen property losses would be noticeably reduced
- Crime prevention success would dramatically increase

Predictive Analytics Influencing Many Insurance Industries

It has been found that many firms across insurance industry have switched to applying predictive analytics to their daily business practices. Notably, 49% of the personal auto insurance industry now uses predictive analytics for:

- Profit increase
- Risk reduction
- Revenue growth
- Overall company operational efficiency improvement

Big data and, specifically, predictive analytics are changing the way insurance companies conduct business and relate to their customers. The trend now is to give their customers the product they want at the price they want. Also, insurance companies have become more sensitive to their customer's core issues. Competition has increased in this key industry, and companies can no longer say, "You do not like our prices? Too bad." Insurance companies can no longer ignore the wishes of their clientele.

If the insured doesn't like the prices a company is offering them, they will go to the competition that will offer them a better insurance rate. How does predictive analytics help in these cases? Well, insurance companies are now offering telemetry-based packages to consumers. These packages are based on predictive analytics. The insurance company uses predictive modeling to form a risk management profile of the customer. They take all the risk factors connected with this particular person in terms of:

- Likelihood of the person getting in an accident
- Likelihood of this person's car to be stolen

By comparing this behavioral information with thousands of other profiles, the insurance company can compile an accurate assessment of the risk factors of this individual. In turn, the insurer can offer the insured a reasonably priced premium. The data is transmitted from a box located in the car or from the person's smartphone app. This process helps insurance companies to bring value to their customers and stay in touch with their needs. The whole point is for the companies to offer better services to their customer base.

One insurance company, U.S. Insurers Progressive, is an example of a company that uses data to enhance their customer services. They have developed what they call "Business Innovation Garage," where their technicians will road test these latest innovations. One example is that they render images of damaged vehicles from computer graphics. The images are sent from cameras and constructed into 3D models showing the condition and extent of damage to the vehicle. This will speed up the claim process. It will also enable the insurance company to process the paperwork quickly. They will have the condition of the car in real time. The customer will get the estimate quicker, and he or she will be able to get the car into the body shop immediately. The customer's car will be back on the road in good working condition in no time. This will definitely help this insurance company increase their customer base.

There are still other methods of conducting predictive analysis that i have not mentioned in this chapter. With these techniques, it is possible to reap the multiple benefits of predictive analytics.

Chapter 11: Introducing R

There are many ways to analyze data, and they all include the use of computers. However, if you are going to really get into the data and perform a good analysis, you will need to learn how to use special software. R is the name of the most commonly used and probably the most powerful piece of data analysis software. R is completely free, and it is a type of open-source software, which means that programmers all over the world are free to make any changes to the code as they see fit.

Unlike much of the applicative software we use on a daily basis, R is command-line oriented, which means that the majority of the tasks you will perform in R will be done by writing lines of complex code, which will tell the software what you want to do. Those who have used Microsoft Excel and thought that is all there is to data analysis will need to guess again as R is a much more powerful tool that also requires a great deal of knowledge to be used properly.

R is a programming language. This means that you will need to understand your data before you can begin to write code to analyze it. The complex syntax and commands of R will take time to learn, but the results you can get from R analysis will be more extensive than anything else you have ever had in the past.

R is simple to download and install from the internet if you are using a PC or Mac, but the software does not work on tablets or smartphones. Once you have it installed, it will take time to start learning how to use it. It will be like traveling into a new dimension.

When you start up R, you will be prompted with an empty screen (barring the about info, etc.) with a command prompt ready for instructions. The command prompt is introduced by the > sign, and any commands you enter at the prompt followed by a stroke of the enter key will send info to R for processing. If all the commands and instructions are typed in the correct format, R will process the information and give you the desired result in the next line. To test out R, you may want to type something along the lines of 1+1 in the command prompt and push enter, and sure enough, R will respond 2, as this is the result of your inquiry. Naturally, R is able to analyze much more complex data than this, and I will talk about some of the possible uses of R in this chapter. So let us briefly discuss R and its potential uses in big data analysis.

Why Use R?

R is an extremely powerful tool for analyzing data, but possibly the most important reason to use R instead of data analysis packages that are so numerous on the internet is that it is a programming language rather than applicative software. This difference will be subtle to a lackey but in reality, it is quite significant. A data analysis package will mainly allow you to run a certain number of operations and perform certain fixed actions. Programming language as extensive as R allows you to specify completely new tasks and run them, which means that you can quite literally invent a new way to process your data and write a code that will do this just the way you imagined it.

If you are not an experienced programmer, R does come with packages as well, and there are over 2000 of them, all of which will allow you to run various types of pre-determined analytics on your data without writing complex code. You will be able to quickly analyze your data, make any type of adjustment to it, and run various ideas and possibilities through the software.

If you are a programmer, then R will be a major thrill for you. It is an open-source programming language, which means that you can not only program in it but also reprogram it all together, thus changing the very code of the language. While this is very complex, many people have done it, and many companies will adjust R based on their needs by changing its source code.

What Type of Analysis can I Run?

This is the best part of R. You can run basically any type of data analysis you want or can imagine. If you have a unique set of data and want to analyze it in a unique way, you can write up a code to do it. If you are looking for common data analysis, you will not be likely to need to write the code yourself, as there will already be packages out there that could do it for you.

I mentioned in the introduction that you can run very simple functions, such as addition or subtraction, with R. Finding the statistical values, such as median or mean, will be equally as simple. No matter how much data you have at hand, you will be able to quickly gain access to these values, which will be crucial for any type of statistical analysis.

From there, a number of tasks you can complete in R will only grow, and some of the most complex data analysis ever completed has been accomplished with R. There is no need to run one function at a time, as complex codes can be made to crunch your data in very meaningful and extensive ways and to provide a host of information on any set of data within minutes.

R or Other Languages?

There are other programming languages, such as Python, that can be used for data analysis as well. While many of these languages are easier to learn and take advantage of, R is the one that will give you the most flexibility when it comes to statistical analysis, and it is widely considered to be the best analytical programming language out there.

Chapter 12: Why Predictive Analytics?

Now that you have a fair idea about predictive analytics, let us now understand why it has gained increased importance over the years. Predictive analytics has found its way into several fields and applications and has had a positive impact so far. In this chapter, I have highlighted the key applications that employ predictive analytics.

Analytical Customer Relationship Management (CRM)

This is one of the most popular applications that employ predictive analytics. Different methods of predictive analytics are used on the customer data available to a company to achieve CRM objectives. CRM aims to create a holistic view of the customer, regardless of where the information about the customer lies. Predictive analytics is also used in CRM for sales, customer service, and marketing campaigns. The different methods of predictive analytics help the company in meeting the requirements of its huge customer base and making them feel satisfied. Some areas where predictive analytics is used in CRM are as follows:

- Analyzing and identifying products of the company that have the maximum demand.

- Analyzing and identifying products of the company that will have increased demand in the future.

- Predicting the buying habits of customers. This will help the company in promoting several of its other products across multiple areas.

- Proactively identifying pain points that may result in the loss of a customer and mitigating such instances.

CRM can be used throughout the lifecycle of a customer, starting from acquisition and leading to relationship growth, retention, and win-back.

Clinical Decision Support Systems

Predictive analytics is extensively used in healthcare for calculating the risk of contracting certain diseases or disorders. It is primarily used to determine the patient's risk of developing health conditions, such as heart disease, asthma, diabetes, and other chronic diseases. If you were under the impression that predictive analytics is used only for diagnostic purposes, then you are wrong. Predictive analytics is also employed by doctors for making decisions regarding patients who are under medical care at a certain point in time. How does it work? Clinical decision support systems aim at linking observations of a certain patient's health with knowledge about health. This relationship will aid clinicians in making the right decisions with respect to the patient's health.

Here are some other ways predictive analysis is helping the healthcare industry:

Predictive analytics is influencing the healthcare industry. It could revolutionize healthcare worldwide. Here are seven ways through which it will impact healthcare:

- Predictive analytics increases medical diagnosis accuracy

- Predictive analytics will support public health and preventive care.
- Predictive analytics gives doctors solutions for their patients.
- Predictive analytics can provide predictions for hospitals and employers about insurance product costs.
- Predictive analytics lets researchers make prediction models without studying thousands of patient cases. The models will become more accurate over time.
- Predictive analytics helps pharmaceutical companies develop the best medicines for the welfare of the public.
- Predictive analytics provides patients with potentially better health outcomes.

These seven points can have industry-changing practices for the healthcare industry. The ability to predict medical conditions with greater accuracy could save lives and medical expenses for patients. Malpractice suits will decrease, and healthcare costs could potentially decrease. Doctors' ability to diagnose illnesses accurately can save patients and insurance companies thousands of dollars in individual cases. Insurance companies could potentially save untold millions in medical claims worldwide.

Imagine doctors diagnosing a patient's sickness right the first time. There would be a domino effect. This should decrease healthcare costs dramatically. People will be able to afford healthcare insurance. Doctors can reduce their rates and live in peace knowing they won't be sued for malpractice. Patients would save money because they wouldn't be spending needlessly on prescriptions that aren't connected to their illnesses. Healthcare would be streamlined in the patient care process if doctors have the ability to use predictive analytics to give patients better care based on accurately researched information.

Elderly patients would be able to afford their medications because, in some cases, they are given many prescriptions that could cost thousands of dollars monthly. Pharmaceutical companies would use predictive analytics to research and manufacture medicines to better meet the needs of the public. It would be possible to reduce the inventory of existing drugs worldwide. Pharmaceutical companies can produce necessary medicines that would directly deal with particular conditions patients may have. By using predictive analytics to find accurate data about patients' health issues, companies can produce the exact medicines that will help cure those issues. If this practice were done on a wide scale, unnecessary medicine thought to help treat conditions but actually don't could eventually be removed from the market. Pharmacies would be left with only the medicines that are needed to treat various diseases.

Accurate diagnosis trend could bring about a huge in change in medical care. Imagine a patient who has a family history of having heart attacks and whose ancestors were prone to having heart conditions. The patient's doctor can use predictive analytics to predict the chances of the patient having a heart attack in the same age range as other family members have had. More to the point, what if the patient is at an age where the risk of heart attack is imminent?

The patient's physician will be aware of all these facts. The doctor can use predictive analytics to predict when the event of a heart attack could occur. Armed with this life-changing knowledge, the doctor can share the information with the patient. The doctor and the patient can develop a health plan to reduce the risk of the heart attack or eliminate the risk altogether. The patient may have extended his or her life by several years because of this health prediction. Better yet, the patient may have saved his or her life in the here and now. These are the kinds of radical changes that predictive analytics can bring to the healthcare industry.

It is reasonable to assume the roles of the doctor and patient will change based on the new health data that predictive analytics has brought and will bring to medicine. The patient, being more aware of care options, will increasingly make better healthcare decisions for him or herself. Doctors will become more of a consultant in advising the patient on the best course of action regarding health choices. It seems as if that trend is already taking place.

Today, more patients are aware of their health issues, and more options are available. Patients are getting more involved in the direct decision-making process about their personal healthcare. Physicians seem to rely more on the patient's awareness about healthcare plans than before.

This is just the beginning of wide and sweeping changes that will occur in the healthcare industry because of the innovation of predictive analytics.

Collection Analysis

There are several industries that have a risk of customers not making payments on time. A classic example of this case would be banks and other financial institutions. In such cases, financial institutions have no choice but to engage the services of collection teams to recover payments from defaulting customers. It is not surprising that there will be certain customers who will never pay despite the efforts of the collection teams. This is nothing but a waste of collection efforts for financial institutions. Where does predictive analytics come into play? Predictive analytics will help the institution in optimizing collection efforts. Predictive analytics plays an important role in collection activities in the following ways:

- Optimizing the allocation of resources for the purpose of collection.
- Identification of highly effective collection agencies.
- Formulation of collection strategies.
- Identifying those customers against whom legal action is required to be taken.
- Formulation of customized collection strategies.

With the help of predictive analytics, it becomes easier for financial institutions to collect their dues in an effective and efficient manner. This also reduces the collection costs for financial institutions.

Cross Sell

This is applicable for organizations that sell multiple products. As previously mentioned, it is important that every organization has the details of its customer base. When an organization has a comprehensive database of customers, predictive analytics will help in the promotion of the other products of the organization. Predictive analytics helps in determining the spending capacity of each customer, their usage, and other behavior, thus making makes it easy for the organization to promote other related products. This will not only lead to an increase in sales for the organization but also help in building customer relationships.

Customer Retention

Customer satisfaction and customer loyalty are two of the major objectives of any organization. These objectives have gained more importance in recent years, given the increase in competition. Apart from these two major goals, another goal has emerged in the light of such heavy competition in the market. This goal is reducing customer attrition. It is not important if new customers purchase an organization's product or services. What is more important is that existing customers continue buying such products or services. When you are capable of retaining your customers, you are bound to increase your profits without much difficulty. Hence, it is highly important that each organization pays attention to the needs of the existing customers.

Most businesses tend to react to the customers' needs rather than proactively addressing them. This attitude could easily break them, should their competitor be more empathetic towards the needs of the customers. Moreover, it is too late to change a customer's decision once he or she has opted to discontinue the service of an organization. Regardless of what an organization says or does, it may not have that much of an impact on the customer. Ultimately, the organization will have to bend over backwards to retain the customer, which is not only an added cost to the company but also an embarrassment.

With predictive analytics, it will be possible for organizations to be more proactive as opposed to being reactive. How does it work? By frequently having a quick glance at the customer's service usage history and spending habits, it will be possible for organizations to come up with predictive models. These predictive models will help in the identification of customers who are most likely to discontinue the service of the organization. This gives the organization an opportunity to act proactively and figure out the reason behind the possible termination of service. It can also decrease the number of customers terminating their service in this fashion. Some strategies used by organizations to retain the customers who are on the verge of leaving include lucrative offers, discounts, etc.

Another kind of customer attrition that is equally detrimental to the interests of a business is silent attrition. This is when a customer reduces his purchase of the organization's products or services gradually over a period of time and then discontinues eventually. Most companies are not aware of customers who have shifted loyalties this way. However, with the help of predictive analytics, it will be possible to pick out customers who are exhibiting this behavior. As previously mentioned, predictive analytics mandates the need to monitor the customer's service history and expenditure regularly. This will help companies in spotting customers who have reduced their spending over a period of time. With this information in hand, it will be possible for the company to come up with a customized marketing strategy for such customers and ensure that they continue purchasing the services of the organization.

Let's introduce some case studies to support the claim that predictive analytics can help increase customer retention and reduce "churning." Churning refers to the case in which a customer stops buying products from a store or business and begins shopping at another retailer's website or physical store.

The first case study is Windsor Circle's Predictive Analytical Suite, which produces many predictive data fields to help vendors in their e-mail marketing campaigns:

- Predicted Order date based on customers personal purchasing habits
- Replenishment Re-Order date, which is when the product they purchased will run out
- Product Recommendation related to products customers have purchased historically
- Hot Combo Products that are typically bought together.

The goal of predictive analytics in direct e-mail marketing is to combine Predicted Order Dates and Replenishment Re-Order Dates with product Recommendations and Hot Combo sales to retain customers by appealing to their predicted spending patterns. The vendors entice customers with future Hot Combo sales by reducing prices and predicting when such items will sell out.

By using Windsor Circle's Predictive Analytics Suite, vendors can create models predicting which customers will buy which products in what particular time frames. Based on historical data showing previous product purchases, vendors can piece together effective direct e-mail marketing strategies to induce customers to keep coming back.

These predictive models can also predict when customers will purchase certain products together and which products they most likely would buy together. Another component of these predictive models is that they can foretell the right timeframes to keep products in inventory based on past customer demand.

Coffee for Less.com uses Windsor Suite's predictive data to run their replenishment e-mails to remind customers to stock up on their coffee purchases. Windsor automatically pulls up a predicted order date on customers who have made three or more purchases. They then trigger replenishment e-mails and send it to the customer. The e-mail is based on the customer's historical buying patterns. In this e-mail example, they would put a smart image to the left. This would then populate a particular product image based on the products the customer was getting the replenishment e-mail about. Shrewdly enough, Coffee for Less will use the rest of the e-mail page to suggest other products the customer will like. This e-mail is generated on predicted products that customers will most likely buy. Here, Coffee for Less is using two data marketing strategies to retain this particular customer. These are Product Recommendations + Predicted Order dates all based on Windsor's predictive analytic models.

Normally, vendors use Windsor's predicted order date field to trigger product recommendations and replenishment e-mails. Australia's premiere surf and apparel retailer, Surf Stitch, used the predicted order date field in an innovative way. They used it to locate potential "churning" customers (defined previously as customers who will stop buying products from a particular vendor). The usual timeframe for win-back e-mails to be sent out is in 60, 90, and 120-day cycles since the last customer purchase.

Surf Stitch has access to their customer's purchase history and their predicted order data. By using the replenishment e-mails in an innovative way (win-backs), they were able to reduce churning customers by a whopping 72% over a six-month period.

The case studies above show that companies (in many industries) can use predictive analytics in an innovative way. They can take standard data fields and use them in new ways to retain their customer base and win back lost customers. Predictive analytics is so flexible in data applications that it can be used in thousands – potentially millions – of ways. There are applications waiting to be discovered that no company has thought about using.

The possibilities that predictive analytics offer for company's business practices are unlimited. Sadly, there are still companies in various industries that are lagging behind in the use of predictive analysis models. They are still using traditional predictive analytic methods that are no longer effective. These old methods are slow, cumbersome, and not cost effective. They take too much manpower and time to facilitate all their processes. They slow down the revenue sources, production, and overall operational efficiency of these companies.

The old predictive analytical models aren't user-friendly and can't predict as many different variables as the new methods can. In the long run, the newer predictive systems will eventually be implemented in all the companies that are currently not using them. What will motivate the change? When the companies see that their traditional predictive analytical models are increasingly costing them their market share to their competitors, they will have to switch. This is because internal processes will not be streamlined and cannot run more efficiently. Their bottom line will not increase but decrease because they aren't using the newer predictive analytical models to their fullest capacity, or they aren't using them at all!

Direct Marketing

For organizations that manufacture consumer products or provide consumer services, marketing is always an important matter of discussion. This is because in marketing, the team should consider not just the pros and cons of their organization's products or services but also the competitor's marketing strategies and products.

With predictive analytics, this job is made easy. It can improve your direct marketing strategies in the following ways:

- Identification of prospective customers.
- Identification of the combination of product versions that are the most effective.
- Identification of effective marketing material.
- Identification of highly effective communication channels.

- Determining the timing of marketing strategies to ensure that such strategies reach the maximum audience.

- Reducing the cost per purchase, calculated by dividing the total marketing cost incurred by the company by the number of orders.

Insurance companies are using predictive analytics to develop personal marketing strategies to attract potential customers. They will analyze many sources of data to find out what customers like. They will analyze customer feedback data, which involves scouring their own websites and social media sites. Algorithms scope through unstructured data in phone calls, e-mails, and information revealed on social media about what customers like or dislike. Moreover, the insurance company will analyze how much time a person will spend on the FAQ section of their website. Even forums and message boards are scanned to see what a customer's preferences are. All of this is done so the insurance company can make a unique profile for the potential customer.

I mentioned in another section of this book that insurance companies are now using all data available to them. They need to focus on each customer's needs individually to stay competitive and, actually, to stay in business. Insurance companies have always focused on whether they are meeting the customer's needs or not in the past. However, they are now doing this more aggressively through the new data analysis techniques available to them. Clearly, they are attempting to determine whether they are having a positive impact on their customers.

The marketing departments of insurance companies are also using predictive analytics to track whether customers will terminate their policies or not. Like other tracing processes, the company will investigate a customer's behavior and compare that behavior to other customer profiles who have actually canceled their insurance policies. This will help determine or predict whether a customer will cancel in the near future. One pattern data source that is analyzed is whether a customer has made a high or low number of calls to the helpline. If the numbers are especially high or low, it will be flagged.

The flagging will alert the company to try and change the customer's mind about canceling his or her policy. They might offer lower premiums or discounts on other services to get the customer to stay, and having this data to make those calls before they cancel is critical. More importantly, they can spend more time solving the customer's issues before they happen. Thus, predictive analytics is playing a crucial role in marketing. Insurance companies are now zeroing in on more customized customer profiles.

In sum, big data is helping the insurance industry improve its customer service, minimize fraud losses, and lower the price of premiums. As a side note, the FBI reports that customers pay $400 to $700 more in premium costs, on average, because of fraud losses.

This is how predictive analytics plays an important role in direct marketing.

Fraud Detection

As previously mentioned, fraud is an inherent threat to every organization. Lack of data makes it difficult for a company to even detect fraud. Fraud can be any of the following types:

- Fraudulent transactions. This is not just restricted to online transactions but also involves offline transactions.
- Identity thefts.
- Inaccurate credit applications.
- False insurance claims.

The size of an organization does not shield it from fraud. Whether it is a small business or a big corporation, it faces the risk of being a victim of fraud. Classic examples of organizations that are often victims of fraud are retail merchants, insurance companies, suppliers, service providers and credit card companies. Predictive analytics can help these kinds of organizations come up with models capable of detecting incorrect data or applicants and thereby reduce the company's risk of being a victim to fraud.

Predictive analytics plays an important role in detecting fraud, both in the private and public sectors. Mark Nigrini developed a risk scoring method that is capable of identifying audit targets. He employed this method to detect fraud in franchisee sales reports associated with an international food chain. Each franchisee is scored 10 predictors at the outset. Weights are then added to these initial scores to give the overall risk score for every franchisee. These scores will help in identifying those franchisees that are more prone to fraud than others.

This approach can be used to identify travel agents who are fraudulent, questionable vendors, and other accounts. There are certain complex models developed using predictive analytics that are capable of submitting monthly reports on the fraud committed. In fact, the Internal Revenue Service of the United States of America makes use of predictive analytics to keep a tab on tax returns and to identify any fraudulent entries or evasion, as well as to catch tax fraud early.

Another growing concern when we speak about fraud is cyber security. Predictive analytics can help you come up with behavior-based models capable of continuously examining actions on a specified network and spotting any deviations or abnormalities.

Widespread fraud is a serious challenge for the insurance industry, but companies are fighting fraud on the front lines through profiling and predictive analytics. The companies will compare current claims with claims that have been historically found to be fraudulent. Certain variables that indicate a claim that is being filed is a fake will be picked up by computer analysis. The fake historical claim has certain markings to show it is a fraudulent document. The claims are matched up, and if the claim in question has the same markings as the historical fraudulent claim, then it will be investigated further. The matches could have something to do with the behavior of the claimant. Criminals who engage in fraud typically display certain patterns that can be detected by computer analysis. These might not get flagged by a human who manually processes these claims, but computers can look at a volume of claims for triggers and alert the appropriate people about the occurrence for further review.

The other factors the company will look at are the people the claimant associates with. These are investigated through open sources, like their activity on social media or the other partners in the claim, such as an auto body shop. The insurance company will be looking at dishonest behavior that the auto body shop may have exhibited in the past. Credit reference agencies are checked out as well.

Operations

Sometimes, analysis can be employed for the purpose of studying the organization, its products or services, its portfolio, and even the economy. This is done to identify potential areas of business. Businesses use these predictive models to manage factory resources, as well as to forecast inventory requirements. For example, airline companies use predictive analytics to decide on the number of tickets that they are required to sell at a specified price range for every flight. Similarly, hotels may use predictive models to arrive at the number of guests they may be expecting during a given night. This way, they can alter their price range accordingly and increase their revenue.

For instance, if the Federal Reserve Board is keen on forecasting the rate of unemployment for the upcoming year, it can be done so using predictive analytics.

Insurance Operations

Today's insurance giants seem to be behind the times in their approach on the use of predictive analytics methods. The current methods require many data analysts to run them. The methods are time consuming and cannot handle the massive amount of data flying through the insurance companies' portals to be effective systems. When you have data funneling through many sources, it makes it improbable for insurance companies to analyze data in a quick and efficient manner. These two factors are a must if the company wants to stay abreast of a highly changing market place.

For companies to be relevant, they must adopt predictive analytic models that will turn business value-specific decisions and actions to maximize operations across the board. Solutions must be continuously improved and refined through thousands of iterations. Analytic output must connect into business operations to optimize revenues and be relevant. Modern analytics provides the following predictive analytic features to insurance companies that want to improve business operations:

- Predictive analytics specifically for insurance
- Predictive modeling for insurance
- Fully operational predictive analytics
- Big data analytics
- Prescriptive analytics
- Customized insight and knowledgeable answers

Insurance companies must realize that proper predictive analysis will improve overall business operations. Unfortunately, there is this mindset within the insurance industry that current predictive models are not efficient. They have worked previously, but in today's market, they've become obsolete and will eventually hurt the overall business operations revenue sources, resource optimization, and sales nurturing of insurance companies.

They need to have an automated predictive model in place to keep up with the influx of data constantly bombarding their systems. Manpower labor is too costly to continue to use manual predictive analytics methods. These resources can be channeled into automated systems to save the companies a lot of time and money. It is true that with the advent of automated predictive models, you would have to hire additional technical support people who understand the complexities of running these automated systems. However, the revenue increase in all the other areas will more than justify the manpower expense. Reluctance on the part of analysts and statisticians to automate predictive processes fully is too costly for insurance companies to continue operating in.

The predictive analytics model that modern analytics offers will be the very model insurance companies need to increase sales, decrease fraud, and increase their policyholder's market instead of losing them. Modern analytics fully automates predictive systems, thus allowing them to run thousands of models at one time. Their methodologies run much faster than traditional analytic systems do. Moreover, they will assist insurance companies in increasing overall productivity, enhancing daily operations, effectively locating strategic business investments, and predicting any changes in the current and future marketplace.

Their predictive analytics models will be game changers for any insurance company that chooses to implement their services. How? By optimizing business operations, improving internal processes, and surpassing competitors in predictive analytics processes. Modern analytics works very closely with their clients, and they have helped a wide base of customers across many industries. They gather and structure data with their cutting-edge algorithms and technology. They also rapidly supply top-notch data solutions uniquely tailored for each client.

Shipping Industry

UPS has adopted a new program called On-Road Integrated Optimization and Navigation, or ORION for short. UPS believes that predictive analytics needs to be implemented all the way down to the front line workforce. UPS is deploying this predictive analytics model to its front-line supervisors and drivers who number 55,000 in total. The predictive model factors in the following variables:

- Employee Work Rules
- Business Rules
- Map Data
- Customer Information

These are used to predict the optimal delivery routes for UPS drivers. The data sources will answer such questions as "Which is better, to make a premium delivery 15 minutes earlier or to save a mile of driving to reach peak performance objectives?" UPS is taking innovative steps to bring their predictive model to execution. Front-line supervisors will learn to interpret predictive analytics data to maximize delivery times on their routes. That is the key behind this move. Within UPS, these data skills are no longer just used by analysts.

UPS originally distributed the ORION system in 2008. The problem was there were too many variables for the managers to manipulate. It confused them because they didn't understand how these variables applied to them. UPS refined ORION, and now, only the essential variables have to be input by managers. The managers teach the drivers how to interpret the ORION predictive data. The data was put into terms the front-liners could understand, and understand they did. One driver, over a few months, cut 30 miles off his route by interpreting the particular predictive data for his route.

Risk Management

Predictive analytics is employed to assess the risk associated with a business in a fashion similar to how it is used to detect fraud. Credit scoring is the crux of ascertaining the risks associated with the business. Credit scores can determine the likelihood of a customer to default in payments. How is this done? The credit score is nothing but a number generated by a predictive model. This model consists of all data associated with the credit worthiness of the customer in question.

Employee Risk

There is another side of risk management that is typically overlooked, that is, the risk associated with a company's employees. Companies can use predictive analysis to gauge employees who may be a risk. There are two traditional methods in place to detect employees who could become troublesome. They are the "blanketed management" programs that focus more on non-risk employees and the employees at risk of slipping by without being noticed. The other method is the "squeaky wheel" approach, which is when company management focuses on employees that display active troublesome behaviors.

Another issue that companies face is how to monitor for employees at risk effectively if they have thousands of employees. They can't use traditional methods. So here is where predictive analytics comes into focus. Predictive analytics will provide specific risk management programs. They use thousands of data points that are turned into tangible data. This data tracks subtle but profound changes in an employee's behavior that could lead to major behavioral issues down the road. Often, these changes go undetected, even with risk management staff employed. Previous trends are also analyzed. These components are interpreted to predict future outcomes of an employee's unusual behavior. This information arms the managers and gives them the ability to intervene in the employee's issues before such issues impact the business.

Dramatically undetected risk could cause employees to engage in tragic behavior. Thus, the early intervention of a manager can sometimes avoid serious negative behavior down the road. Managers can intervene with the right person, at the right time, and on the right subject. These interventions can stop worker compensation claims and voluntary employee turnover.

Their blanket management methods are replaced with these streamlined task-specific models that will save time, money, and, most importantly, reduce risk. The value of these predictive models is that they can historically analyze employees' past behaviors to identify present and future behavioral patterns.

I need to point out here that when employees are hired in the beginning, they aren't at risk. The events that develop after their hiring are the ones that put them at risk. Problems in the employees' lives can surface outside of the workplace. They could be having medical or personal issues that carry over to their performance at work. Again, as stated earlier, predictive analytics will spot subtle changes in an employee's behavior from a historical analysis of records.

Let's briefly talk about the transportation industry in relation to risk management. This industry is highly vulnerable to risk with its employees. The transportation industry uses predictive analytics to track the unsafe driving patterns of employees. The data reveals that an employee may be driving unsafely, which could lead to violations, fines, and possible costly losses due to accidents. People may even be killed in an auto or other type of vehicular accident. Specifically, one type of hazardous driving pattern could be one in which an employee does not pay attention to his or her driving because he or she has a sick, elderly parent. This driver is too busy talking on the phone while driving, setting up doctor's appointments and other activities for this sick parent.

Based on historical data patterns, the transportation manager can detect whether the employee is driving abnormally. He or she could be braking extra hard or spending too much time in idle. The manager can communicate with the employee in question to work out some type of solution to help the employee deal with issues to avoid a costly accident. A solution could be a reduced workload or adjusted schedule.

Predictive analytics can help a manager focus on why an incident happened instead of what happened. Managers tend to focus on what happened instead of why it happened. The managers' focus is directed on correcting the issue and not necessarily the root cause. Changing focus on the incident can solve employees' personal issues much quicker.

Underwriting

Many businesses will have to account for their exposure to risk. This is because of the different services offered by the organization. They will have to estimate the cost associated with covering these risks. For instance, insurance companies that provide insurance for automobiles need to determine the exact premium to be charged for the purpose of covering every automobile and driver. In the case of a financial company, they will need to analyze the potential and the repaying capacity of a borrower before granting a loan. In the case of a health insurance provider, predictive analytics can be employed to figure out the following information: past medical bills of the person, pharmacy records, etc. This information can be used to predict the quantum of bills that the person may submit in the years to come. In this way, the health insurance provider can arrive at a suitable plan and premium.

Predictive analytics can help you underwrite these risks. With the help of predictive analytics, it is possible to determine the risk behavior of a certain customer when he or she makes an application. This can help companies make a calculated decision.

Predictive analytics has also reduced the turnaround time for processing loans and other claims based on credit scores. The reduced turnaround time has made a big difference in the case of mortgages. These decisions are now made in a matter of a few hours. This is different from the erstwhile procedure, whereby it would take weeks for financial institutions to offer a mortgage. Predictive analytics also helps in the pricing of financial products, which can reduce the chances of default. Quite often, the most common reason for default is the fact that the interest rates or the premiums are too high.

Recently, the insurance industry has turned to predictive analysis to help solve the issues they face. They are using this technology to analyze credit scores to foretell loss-ratio performance and the behavior of the insured. Now, insurance carriers are using predictive analytics to see into the future during underwriting. They are creating models to see how future policies will perform. They face a very serious issue in which they really have no reliable data on the homes they insure. There is a big risk that the houses they insure could decrease in value in the market. What if the house burns down? Is the house vulnerable to earthquake damage? Is the house located in a potential flood zone? These are the types of questions that predictive analysis will help them answer.

With unknown data about the houses they insure, they are at risk of losing millions of dollars in natural disaster damage or weather damage. There is also the insurance/deficiency value and the potential for liability risk. Here is where predictive analysis comes into play. Predictive analysis is cheaper, faster, and more efficient than traditional analysis methods. Time will tell if predictive analysis will be a real game changer or not.

Predictive Analytics: A Case Study in Underwriting – Freedom Specialty Insurance

In the aftermath of the major economic slowdown of 2008, a risky underwriting proposal was submitted to the executive management team of Scottsdale Insurance Company. This case study is taken from the D&O insurance industry (D&O refers to Directors and Officers Liability Insurance. This liability insurance is paid to the directors and officers or the organization itself. The money is reimbursement for losses or a loan of defense fees in case an insured experiences a loss because of a legal settlement.).

The Scottsdale Insurance Company accepted the proposal, and Freedom Specialty Insurance Company was born. Freedom developed an industry-first method, feeding external data into a predictive model to support risk intuition. The theory was that class action lawsuit data could predict D&O claims. A multi-million-dollar unique underwriting platform was born. The dividends are beginning to pay off as Freedom has grown to $300 million in annual direct written premiums. They have consistently kept their losses under control. In 2012, they were under the average industry loss rate of 49%.

Their model delivers a high level of honesty in their dealings. This has caused great trust among personnel, from the underwriter to the executive branch in the parent company. The reinsured are confident, as well. The point of this case study is to show how Freedom built, maintained, and refined their underwriting model in hopes that other companies can take this model and incorporate it into their underwriting operations by using lessons Freedom learned along the way. The platform was a multi-team effort in its development. An actuarial firm built and tested the predictive model. The external technology supplier built the user interface and constructed the integration with corporate systems. Next, technology from SAS was used for many parts, such as data repositories, statistical analytics engines, and reporting and visualization tools.

Their platform involves the following components:

- Data Sources
- Data Scrubbing
- Predictive Model
- Risk selection Analysis
- Interface with corporate systems

These are described as follows:

Data Sources – The system integrates with six external data sources and receives data from corporate administrative applications. External sources, like class action lawsuit and financial information, are commonly used in the D&O industry. Other sources are specific to Freedom and drive their predictive model. Data is central to the platform's function, so Freedom spends a lot of time maintaining quality data and vendor activities. The back-testing and categorization processes illuminate vendor weaknesses and incongruities in vendor data. Freedom goes to great lengths to work with vendors to classify data and keep its quality high.

Freedom has realized that they need to keep an eye on their eternal data as well. They apply strict checks to improve policy and claims data. Freedom absolutely maintains data freedom from vendor's identification schemes. It takes extra work to translate values, but this ensures that Freedom can get rid of vendors quickly if they have to.

Data Scrubbing – The data goes through many cleansing cycles when it's received. This guarantees maximum usefulness. One example of this is the review of 20,000 individual class action lawsuits monthly. Initially, they were categorized by different parameters, but they are reviewed monthly to see whether any changes have taken place. In the beginning, this process took weeks to complete when done manually. Now, it's done in hours by using advanced data categorizing tools.

Back-testing – This key process calculates the risk involved when an initial claim is received. The system will funnel the claim back through the predictive model, testing the selection criterion and changing tolerances as needed. The positive feedback loop refines the system through many uses.

Predictive model – Data is condensed and fed through a model, which uses multivariate analysis to determine the best range of pricing and limits. Algorithms evaluate the submission across many predetermined thresholds.

Risk Selection Analysis – This produces a one-page analytical report of suggestions for the underwriter. Comparisons with the same risks are displayed along many risk factors, including industry, size, financial structure, and other parameters. The underlying principle of the platform is that it's driven by underwriter logic, helped by technology. It will never replace the human underwriter.

Interface with corporate systems – Upon a decision being made, selected information is sent over to corporate administration. The policy issuance process is still largely done manually. There is room for complete automation in the future. The policy is issued, and the statistical information is looped back through the data source component. As claims are filed through loss, data is added to the platform.

Like any D&O operation, underwriters possess deep technical insurance knowledge. They have had to adjust to and become comfortable with a system that reduces mass amounts of information into several analytical pages, because of which underwriters traditionally spent time collecting, collating, and analyzing data. They now interface with policyholders and brokers in book and risk management. Again, predictive analytics has streamlined the data process by changing the traditional role of the underwriter.

Of course, with this revised underwriting process in place, skilled technical personnel were added. These people have legal and statistical knowledge that enable them to build predictive models in the financial field.

Incorporating this model has allowed Freedom to develop process expertise in many areas. The data management tasks, such as data scrubbing, back-testing, and categorization, were learned from scratch. Initially, these processes were handled manually, but they have been continuously automated since their inception. In addition, the number of external sources is always expanding. Freedom is in the process of evaluating the use of cyber security and intellectual property lawsuits. The predictive model is constantly undergoing refinement. The tasks involved in the care, feeding, and maintenance of the predictive model used in other industries have been developed in the D&O industry. Of special interest is that after the actuarial initially built the predictive model, it took many months for Freedom to gain a full understanding of its intricate operations.

Processes were put in place to effectively manage all these external vendors together. Several external groups collaborated to pull the predictive model project together, with each group having to work in tandem with the other groups. Below is a list of those groups:

- Actuarial Firm
- IT Firm
- Data vendors
- Reinsurers
- Internal IT

I mention the groups involved in this project again because it took a very skilled and concerted effort by Freedom to bring these entities together to work with their skills in the same place, at the same time.

Success of the Platform

This predictive analytics model was very successful for Freedom. It has opened up new horizons for the insurance company. First, the analysis was so specific that it now opens up dialog with brokers and policyholders in risk management discussions. Second, this model may be extended beyond liability lines to property lines. Expanding this model will make it applicable to other lines of liability, such as surety. This predictive analytical model makes Freedom stand out from its competitors. Finally, the methods used in the back-testing and categorization elements can be employed to forecast other data elements. Freedom has positioned itself to discover other potential information sources.

Predictive Analytics is Impacting Real Estate

How can Predictive Analytics impact Real Estate?

One example is a real estate corporation that used data devices to help a law firm decide on whether it needed to move into a different office space. They started out by addressing employee retention factors. The law firm wanted to recruit and maintain the best prospects. As this was the major concern for the client, the real estate firm took this initiative instead of approaching it based on a direct real estate suggestion. The firm used various location-aware devices to track the movements of employees. The tracking data was employed in data mapping based on employee preferences and movements.

The interpretation of the data resulted in the law firm moving out of the high-rise office space into a cheaper office space. The data saved the law firm money and positioned them in a new location. The improved location solved the employee retention problem.

Predictive Analytics is changing the way of the National Association of Realtors (NAR)

Predictive analytics is helping the NAR re-think the way they conduct business. They are America's largest trade association at 1 million members. So their results obtained through predictive analytics could influence the real estate industry dramatically. The NAR is looking at how to add value to their client relationships. They established a new analytics group that will:

- Analyze member and customer data trends
- Help NAR use the data results to add value to its realtors
- Use disparate data models to build analytical models
- Employ models to solve complex problems in the housing industry
- Help real estate agents, state and local associations, and others make better data-driven decisions

The data group will step forward in three phases. The first phase is the experimentation phase, where the data group will interpret the accumulative data and identify trends and patterns. The last two phases will involve forming partnerships with other data users; these partnerships will design and develop products to help their agents. The golden result of applying predictive data models will help the NAR uncover behavioral patterns and build targeted meetings with potential home buyers. This will result in enhanced operational processes for the NAR. Ultimately, the NAR will increase their financial bottom line.

The NAR's traditional process of analyzing data is to gather holistic big picture proprietary data. The extracted data shows what had happened historically in the group's business dealings. The other part of analyzing data is gleaning information from the customer relationship management system to determine current events that are affecting the group.

The new and improved big data analysis process will focus on the big picture and the finer details in one source. The new process will analyze both proprietary and public data systems to discern trends that will predict the future in real estate business practices.

On the other side of the coin, insurance companies will be changing the way they conduct business as well. As predictive analytics helps doctors and patients consult to create better and faster healthcare solutions, patients will stay healthier longer and will heal faster because the right treatment options are being administered to them to meet their physical needs. Insurance companies will have reduced premiums and claims to pay on. Yes, the insurance companies will lose millions of dollars in premium costs because patients won't need extensive and unnecessary medical procedures. This will surely drive down health insurance costs.

However, insurance companies can increase revenues in new ways, as the medical profession becomes more focused and singular in its individual patient healthcare plans. This will be reflected in specialized but shorter hospital admissions. Insurance companies may become more specialized in what they cover. The result will be new avenues of revenue.

Medical device companies will be impacted as well. The

familiar medical devices they have been manufacturing to take care of traditional healthcare issues will diminish. The obvious conclusion would be lost revenues for these companies. Nevertheless, the device companies can adjust to the new playing field. They can shift gears and start manufacturing specialized devices that will undoubtedly be needed in the new healthcare environment, thereby bringing increased revenues to their financial bottom line as well. Predictive analytics is bringing widespread change to every facet of the healthcare system. More unseen changes are yet to come as predictive analytics processes become more refined and accurate.

Predictive Analysis is Changing Industries

It is obvious that predictive analysis is changing the game drastically in terms of how data is collected, interpreted, and predicted. The section on "Why Predictive Analytics" shows that predictive analytics is changing the face of so many industries internationally. Many industries are on the cutting edge of the advancements predictive analytics is making in various applications and in specific phases of those industries. Freedom Insurance, which has developed a predictive model that is spreading to other types of applications in the insurance industry, has placed themselves as the leaders in their industry.

Predictive analysis is reported to be a very lucrative market, reaching 5.2 billion to 6.5 billion by 2018/2019. Notably, not millions but billions will be spent in this market.

Take the healthcare industry, for example. Predictive analytics is changing the way physicians take care of their patients. Traditionally, the doctor would just give the patient a medical treatment regimen to follow so the patient could recover and return to his or her full health. Now, physicians have become more like counselors who advise patients on the best course of treatment to take. The patient is intricately involved in his or her own care now because of predictive analytics. The patient has access to his or her medical records through his or her personal physician.

Gone are the days when the patient was unaware of his or her medical history or vulnerability to different diseases because of genetic factors and family history. Now, because of predictive analytics, patients view their records and witness the multiple predictive models that will steer them into the best care for their future health. They now have a strong influence in which predictive treatment will best bring about the desired result – good health. Here, the roles of doctors and their patients are gravitating towards definite role changes. Who would have thought five to 10 years ago that predictive analytics would change healthcare so much? (Maybe, someone somewhere created a predictive model that anticipated the different changes that are affecting the health industry now).

In all the research conducted for this book concerning predictive analytics, every time the predictive analysis applications have been used, they have improved the processes of companies. Every single case studied for predictive analytics has shown that these newer models can help industries improve their overall operational capabilities. Some of the industries that have been radically changed by predictive analytics are as follows:

- Law enforcement (crime scene reconstruction)

- Insurance (risk management concerning employees and customers)

- Transportation (predicting troublesome behavior and averting thousands of citations, arrests, and millions of dollars lost in property damage due to accidents)

- Healthcare (This is a big one because doctors can now effectively diagnose illnesses, conditions, and surgeries based on predictive models. Patient and physician roles are overlapping and integrating together. It's more of a democratic process where the patient and doctor jointly decide the best treatment options. Many more lives will be saved because of updated and accurate healthcare treatments based on healthcare predictive models.)

- Retail Industry (Predictive analytics is helping retailers increase customer retention rates, in some cases by large percentages.)

- Real Estate (The NAR, for example, is using predictive analytics to predict what preferential property packages consumers will likely buy according to historical home buying practices. Rental property investors are using predictive analytics to predict where the most lucrative rental properties may be found based on renter's economic, geographical, and family needs.)

It's not unrealistic to say that predictive analytics has caused industrial revolutions in some industries. The potential market will exponentially increase in the next few years, as more and more success stories are detected relating to innovative ways predictive analytics is being used by visionary companies.

Chapter 13: From Descriptive to Predictive Analysis

Having a powerful business intelligence department is something that every company will benefit greatly from. People who are able to collect and successfully describe a large amount of data will be a huge plus for any company, and this intelligence can be crucial to your business. Still, intelligence alone is not sufficient, and you will need to be able to understand the data described and use it to make predictions through predictive analysis methods.

Descriptive analytics looks into a business on the macro level and examines large amounts of data to find values that may be of use in future analysis. This future analysis is called predictive analysis, and unlike descriptive analysis, predictive analysis builds analytic models on the very lowest levels of business. These models include individual customer or product reports that help with making decisions for future business by anticipating the behaviors of customers and employees, rises and falls in the market, and other similar data using the information gathered through descriptive analytics.

If you look at the answers you can find based on these two types of analysis, you will be able to answer the question of what the difference is between descriptive and predictive analysis.

Descriptive analysis allows you to ask various questions about the history of your business dealings. These questions may have to do with the demographics of your customers, the value of your particular stores, or the sales of particular products. Questions you may find answers to using descriptive analytics are the likes of "What product was our best selling product last month?" "What is our customers' average yearly salary?" "How much money do customers spend on products in our niche per year?" These types of questions will always give us answers about the past and are based on hindsight.

On the other hand, predictive analysis gives us answers about the future of the business. The questions you may want to ask with predictive analysis are often much more useful than those with descriptive analysis. You will be asking such predictive questions as "What is the best price we can put on our product?" "What promotions do our customers want to see?" "How likely are our customers to buy this product?" These types of questions will provide answers that can immensely aid in your business growth and allow you to focus your energy and time on things that will raise revenues.

This move from retrospective to future prediction is very significant, and it is what makes predictive analysis more valuable than descriptive analysis alone. One cannot exist without the other, though, so these two are interlinked in an unbreakable chain. Without descriptive statistics, it is impossible to make any predictions, and without making these predictions, the descriptive data will be next to useless.

Chapter 14: Critical Success Factors

As previously stated, more and more companies have been getting into big data analysis as a part of their business endeavors. These business intelligence projects promise fantastic results and increases in revenue, so quite naturally, all the companies that wish to progress have at least thought of investing money into such business intelligence.

Nevertheless, not all is perfect with big data at this time. While in theory, every company can benefit from it, there have been more than a few cases where companies have been forced to abandon projects, or the results of such projects were much less accurate than expected. In fact, in many cases, the analysis provided data that was completely wrong and useless to the companies.

A study performed in 2007 showed that 34% of all business intelligence projects were either late or too expensive in relation to the budget that was allocated to them. Notably, 31% were abandoned by companies, whereas 17% of such projects went so badly that the existence of their respective companies was brought into question. While some companies, such as Continental Airlines, have made 1000% ROI on their business intelligence investments by being able to lower their overall costs significantly, other companies have often either seen only slight ROI or even losses.

The reason so many companies fail in their business intelligence efforts may actually stem from the fact that very little real research has been conducted into what the critical success factors of a business intelligence project are. A successful business intelligence project should, in theory, be able to integrate a large amount of data into their system and acquire the analytical capabilities necessary to analyze this data in a timely and correct manner. The following measures may be crucial to determining the success of a business intelligence project:

- Data Integration
- Quality of Information Content
- Quality of Information Access
- Analytical Capabilities
- Use of Data in Business Practices
- Analytical Decision Making

The following question must be asked: What are the critical success factors that will lead to the success of a business intelligence project? According to various sources, the critical success factors of business intelligence projects are related to a number of key dimensions, namely, process performance, organizational factors, technological factors, and infrastructure performance.

There are many critical success factors that fall into these categories. These include communication, project management, vendor and customer relationships, user participation, etc. Given that it is impossible for companies to focus fully on all of these, it is necessary to learn what the most important of these success factors are and focus the majority of data analysts' efforts on these significant factors.

According to most experts, the top five critical success factors for the success of a business intelligence project are:

- Top management support
- Flexible technical framework
- Change management
- Strong IT and BI governance
- Aligned BI strategy and business strategy

Top Management Support

As with many things in business, it is very difficult to run a successful business intelligence project without the support from the very top. It is the senior managers who need to have a vision for the BI department and allocate the necessary resources so that the project could be successful. Continued support from the very top of the company has often been cited as one of the key factors in such projects as, without it, it is nearly impossible to get the needed funding, human resources, and other required resources.

Flexible Technical Framework

When the business intelligence system is being put in place, the initial establishment of the hardware and software necessary for the analysis of large amounts of data that will be able to expand and remain flexible as the project grows is a time-consuming and expensive step. However, having more time and money put into this part of the project will save time and resources at a later time.

It is very common to hear that a company's data warehouse stops working properly after some time, and the most common underlying reason is the fact that the technical framework that was established at the start is inadequate. It is necessary for management to invest enough resources at the onset of the project to handle all the data that may become available at a later point.

Change Management

Change is always a good thing as it provides new opportunities and opens new horizons. When business intelligence projects are put into place, it is often necessary to make changes in the way things are done and, sometimes, even in the people who run things. Encouraging your employees to adopt new and advanced ways of doing things is important for every company that wants to have a successful business intelligence system.

Strong IT and BI Governance

Among the main reasons for the failure of business intelligence efforts is the poor quality of the original data delivered to the analysts. This data is collected on the operational side, and it is crucial that systems are put in place that ensure the quality of the data collected for the BI staff to conduct a good analysis of the data. An IBM funded survey found that companies who have good information governance strategy are far more successful in their business intelligence governance as well.

Aligned BI and Business Strategy

For business intelligence to be successful, it is important that the general business strategy and business intelligence are aligned. In many cases, the exact reason why business intelligence incentives fail is because they do not align properly with the business plans of the company. If the BI sector is not fully instructed with a clear business strategy, it is extremely difficult for them to fully leverage big data analysis as a means of increasing overall revenue.

Chapter 15: What to Expect

When we look at the promises big data brings for businesses in terms of social media, we can see that big data may indeed become the factor around which all companies build their businesses at some point in the future. However, big data is not yet at a level where it is accessible to every company, and the costs in both monetary and human resource terms often outweigh the benefits of big data projects.

Many companies and people in the industry have been working on finding the best practices to apply to all big data projects, but at this moment, most agree that there are still no fixed best practices but rather a set of promising emerging practices that may or may not prove useful to different businesses.

While some companies are waiting for better practices to be developed, others are investing billions of dollars into massive big data experiments and possibly gambling away a significant part of their companies. Nevertheless, if successful, business intelligence projects can pay huge dividends, making it a perfect field for risk takers.

Technological Advancements

In terms of technologies used for business intelligence projects, the development of software such as the programming language R or initiatives like AMPlab that combine the force of algorithms, machines, and people are all significantly simplifying the process and making it more accessible to the masses.

The development of GraphChi back in 2012 simplified the process of big data analysis to a level where it can be done on a single PC instead of on supercomputers. It quite literally enabled private users to analyze big data in their homes. This program, for instance, managed to analyze a database of 40 million users and 1.5 billion social media posts on a single PC in 59 minutes. The same task was previously done in 6.5 hours by 1000 powerful computers combined, which means that GraphChi significantly revolutionized the way we analyze data.

Hypertargeting

A simple question that one can always ask about big data is how significant it really is. Let's say we have gathered tons of data; what do we do with it now? Jeff Dachis of the Dachis Group put it quite well, explaining how our entire world today is connected through social media, and with so much information being shared online, brands can quite easily turn this into hundreds of billions of dollars in brand engagement value.

With hundreds of millions of accounts on social networks, such as Facebook, Twitter, and LinkedIn, it is easy to see why companies are investing millions of dollars in engagement with these users through social networks. By gathering all the data and connecting directly with the audience, they are able to hyper target very specific parts of the population and reach their customers in ways that were never before imaginable. Tools like GraphChi now allow us to track such engagement processes and establish very tangible numbers, such as ROI.

Can Big Data Get Out of Hand?

Big social data has become one of the favorite things for companies to collect. It includes all types of social behaviors on networks, such as what we are tweeting when we are publishing posts and how many we publish, etc. Companies use such data to determine what we like and dislike and what we are buying and want to buy, thus potentially increasing their revenues.

Some experts, however, warn us that just like customer relationship management, which was all the rage back in the 90s, turned into an extremely expensive and frustrating practice, the same may happen with big data in the future if it is not approached in the right way. Whether big data will turn into a gold mine or a pitfall for major companies remains to be seen.

Predicting Without the Data

According to some experts, we are, in fact, past the golden time in predicting human behavior. Some believe this was easiest to do about five decades ago, when consumer information was not easily accessible, but few simple factors, such as frequency and monetary value, were able to turn direct marketing into a huge success. Knowing how recently the last purchase was made by a consumer was apparently pretty much the only thing you needed to make them repeat this process by directly marketing more products to them.

Complete Data Management: The Five Major Social Sources

As technology advances, an increasing number of data sources become available on a daily basis. While in the past, we could only track the traditional enterprise data, there are now many other "social" sources that we may track to gather data on customers. These sources include search engine traffic, sensor data, social media data, and semantic data.

These sources are each unique in their own way. While one tells us what people are searching on the internet, another can tell us what kind of buying patterns they are using. While some are internal to a company, others are completely external, such as search engine data. While having only one set of data available may be helpful, gaining access to data from all these sources and analyzing it in combination provides us with complete data management, which allows us to fully understand and better predict the future behavior of customers.

- Sensor Data: This category of data includes data gathered from smart meters that track the energy consumption of households or data collected at the cash register of a supermarket on the purchases made. This kind of data helps optimize processes in various segments of life.

- Social Media Data: The "social listening tools" are used to gather data from social media posts. When we post on Facebook or Twitter, there is a huge amount of data we are letting companies know about us, and make no mistake, all of this data is being collected by various entities and is being used in many ways, from marketing to branding purposes, by these companies.

- Search Engine Data: Tools like Google Trends are used by companies as external sources of data, which allow them to make all sorts of inferences. Google allows their customers to get very deep insights into what we are searching on the internet, and this type of data is being used for search engine optimization and for the targeting of particular segments of the population.

- Enterprise Application Data: This type of data can be used to track purchasing behaviors and other social patterns. This kind of data is usually owned by a company and collected in their own CRM or other systems. This data may include web analytics of the company's own website, and such data serves the building blocks of the entire business intelligence initiative for most companies.

- Mobile Data: Much like using social media, when you use your mobile apps, you are also dishing out information. This data is being collected by such tools as App Analytics and is then sold to companies. Such data may include your geolocation, gender, or age.

It is only by combining these five elements of big data analysis that a company can have full insight into the market as a whole. This is why most analytics vendors are attempting to successfully apply a combination of all these crucial factors into their packages and why all the big companies are doing their best to access and fully utilize each of these five elements.

Chapter 16: What is Data Science?

Data science is defined as "deep knowledge discovery through data inference and exploration." Data scientists need to have experience with algorithmic and mathematical techniques so that they can solve some of the most complex problems by using pools of raw information to find the insights hidden beneath the surface. Data science is centered on analytical rigor and is based on evidence and on building decision-making capabilities.

Data science is important because it allows companies to operate more intelligently and to strategize better. It is about adding significant value by learning from data. A data scientist may be involved in a wide variety of different projects, including:

- Tactical optimization – improving business processes, marketing campaigns, etc.
- Predictive analytics – anticipating future demand, etc.
- Nuanced learning – developing a deep understanding of consumer behavior
- Recommendation engines – making recommendations for Netflix movies or Amazon products, etc.
- Automated decision engines – self-driving cars, automated fraud detection, etc.

While the objective of all of these may be clear, the problems that arise require extensive expertise to solve them. It may be that a number of models, such as predictive, attribution, segmentation, etc., need to be built, and this requires extensive knowledge of machine-learning algorithms and very sharp technical ability. These are not skills that you can pick up in a couple of days; they can take years. Below, we look at the skill set required to become a data scientist.

Data Science Skill Set

Data science is a multidisciplinary job, and there are three main competencies required.

Mathematics

The very core of determining meaning from data is the ability to be able to see such data in a quantitative way. Data contains patterns, textures, correlations, and dimensions that are expressed numerically. Determining any meaning becomes a kind of brainteaser requiring mathematical techniques to solve it. Finding the solution to any number of business models will often require that analytic models be built. Such models are grounded in the theory of hard math. It is just as important to understand how the models work as it is to understand the process of building them.

One of the biggest misconceptions about data science is that it is all about statistics. While this may be important, statistics is not the only math that has to be understood by a scientist. There are two branches of statistics, namely, classical and Bayesian. Most people who talk about statistics are talking about the classical type, but a data scientist needs to understand both types. Moreover, they need to have a deep understanding of linear algebra and matrix mathematics. In short, a data scientist has to have very wide and deep knowledge of math.

Hacking and Technology

Before we go any further, let me just clarify something, I am not talking about breaking into computers and stealing sensitive data when I discuss hacking. I am talking about the ingenuity and creativity required for using learned technical skills to build models and then find the correct and clever solution to a problem.

The ability to hack is vital because a data scientist needs to be able to leverage technology to acquire vast data sets and to work with algorithms that, for the most part, are complex. Just being able to use Excel is not going to cut it in the world of a data scientist. They need to be able to use tools like R, SAS, and SQL, and for that, they need to have the ability to code. With these tools, a data scientist is able to piece together data and information that are not structured and bring out the insights that would otherwise remain hidden.

Hackers are also algorithmic thinkers – they are able to break down a messy problem and turn it into something that can be solved. This is a vital skill for data scientists, especially as they work very closely with algorithmic frameworks that already exist, as well as building their own, in order to solve an otherwise complex problem.

Business Acumen

One of the most important things to recognize is that a data scientist is a strategy consultant before anything else. Data scientists are valuable resources in companies because they and they alone are in the position to be able to add significant value to the business. However, this means that they have to know how to approach a business problem and how to dissect it, and this is just as important as knowing how to approach an algorithmic problem. Ultimately, value doesn't come from a number; it comes from the strategic thinking that is based on that number. In addition, one of the core competences of a data scientist is the ability to tell a story using data. This means that they have to be able to present a narrative that contains the problem and the solution, derived from insights gained from the data analysis.

What is a Data Scientist?

One of the defining traits of a data scientist is the ability to think deeply, coupled with an intense curiosity. Data science is about being nosy, asking questions, finding new things, and learning. Ask any true data scientist what the driving factor is in his or her job, and none will tell you that it is money. Instead, data scientists will tell you that it is all about being able to employ creativity while using ingenuity to solve problems and to be able to indulge curiosity on a constant basis. Finding meaning in data is not just about getting the answer; it is about uncovering what is hidden. Solving problems is not a task; it is a journey – an intellectually stimulating one that takes them to the solution. Data scientists are passionate about their work, and they derive great satisfaction from meeting a challenge head on.

How Analytics and Machine Learning are linked to Data Science

Analytics is now one of the most-used words in business talk, and while it is used quite loosely in some cases, it is meant as a way of describing critical thinking of a quantitative nature. Technically, analytics is defined as the "science of analysis" or, in easier terms, the process of making decisions based on information gained from data.

The word "analyst" is somewhat ambiguous as it covers a range of roles, such as operations analyst, market analyst, financial analyst, etc. Are analysts and data scientists the same? Not quite, but it is fair to say that any analysts are data scientists at heart and in training. The following are a couple of examples of how an analyst can grow to be a data scientist:

- An analyst who is a master at Excel learns how to use R and SQL to get into raw warehouse data.
- An analyst who has enough knowledge of stats to report on the results of an A/B test goes ahead and learns the expertise needed to build predictive models with cross validation and latent variably analysis.

The point I am trying to make is that moving from being an analyst to a data scientist requires a great deal of motivation. You have to want to learn a lot of new skills. Many organizations have found a great deal of success in cultivating their own data scientists by providing the necessary resources and training to their analysts.

Machine learning is a term that is always used when we talk about data science. Put simply, machine learning is the art of training algorithms or systems to gain insight from a set of data. The types of machine learning are wide-ranging, from a regression model to neural nets, but they all center on one thing, that is, teaching the computer to recognize patterns and recognize them well. Examples include:

- Predictive models that are able to anticipate the behavior of a user
- Clustering algorithms that can mine and find natural similarities between customers
- Classification models that can recognize spam and filter it out
- Recommendation engines that can learn, at an individual level, about preferences
- Neural nets that learn what a pattern looks like

Data scientists and machine learning are tied closely together. The scientist will use machine learning to build algorithms that are able to automate some elements of problem solving, which is vital for complex projects that are data-driven.

Data Munging

Raw data is often very messy and has no real structure, and data munging is a term that we use to describe the process of cleaning such data so that it can be analyzed and used in machine learning algorithms. Data munging requires very clever skills in hacking and the ability to recognize patterns so that vast amounts of raw information can be merged and then transformed. Dirty data hides the truth that may be hidden beneath the surface, and if it isn't cleaned, it can be misleading. As such, a data scientist has to be good at data munging so that he or she can have accurate data to work with.

Chapter 17: Further Analysis of a Data Scientist's Skills

One of the best people from whom to seek a worthwhile opinion regarding the skills necessary for a data scientist is someone whose job is to hire data scientists. The reason is that recruiters know precisely what skills they are looking for in the potential employee or consultant. Burtch Works is one such firm that deals with the recruitment of senior personnel in the field of business and industry. They have some work that was published in 2014, and it spells out the skills a data scientist should have. Such information is credible, not just because it comes from a renowned firm of experts, but also because the firm itself has managed to walk the walk and climbed up the ladder of success to join the Fortune 50.

Why don't we borrow a page from Burtch Works and implement their recommendations as regards equipping potential data scientists with relevant skills?

Further Skills Recommend for a Data Scientist

1) Solid educational background

From research, it is evident that people who manage to specialize as data scientists have a solid educational background. In fact, a respectable 88% of them have a Master's degree, a good 46% are Ph.D. holders, while others are generally well-educated. Being a data scientist demands that the person is capable of developing the required depth of knowledge, and one can only do that with a good level of education.

Yet, even with a good level of education, a potential data scientist needs to have an inclination towards fields that declare you are not afraid to do calculations, that is, fields that deal with analysis of numbers and formulas and so on. According to Burtch Works, 32% of data scientists have a great background in Mathematics as well as Statistics, 19% in Computer Science, and 16% in Engineering.

2) Competent in SAS plus/or R

SAS stands for Statistical Analysis Software, and it is only natural that a potential data scientist should be comfortable using software that helps in data analytics at an advanced level, such as data management, predictive analytics, and such. It is even better if the person knows something about the R programming language, which is helpful in creating important functions. Generally speaking, any good analytical tool available is good for a data scientist.

3) Skills in Python Coding

According to Burtch Works, employers who seek to employ data scientists want someone who has technical skills in the use of Python, which is a popular coding language. Often, they are thrilled if the person is comfortable with the use of Java, Perl, or even C/C++.

4) Knowledge of Hadoop

It is important that a data scientist be skilled in the use of the Hadoop platform. These technical skills may not be mandatory, but it helps to be able to derive statistical data with ease from such an open-source library. Other technical skills that are preferred include competence in the use of the Apache Hive – a type of data warehouses software. The Hive helps in querying data by using a language almost similar to SQL, which goes by the name HiveQL.

Whoever is skilled in the use of Apache Pig on top of Hadoop stands a good chance of securing a job as a data scientist. Apache Pig is another platform that helps in data analysis. Also, as long as a person is interested in making a career as a data scientist, it is advisable to become familiar with existing cloud tools, such as Amazon S3, and any other tool that may be developed over time.

5) Skills in SQL Database

It is important that a potential data scientist be able to work with Structured Query Language (SQL), which is a programming language used in data management and stream processing. A data scientist needs to be skilled in writing complex SQL queries, as well as in executing them. It is helpful for data analysts to know tools that will help then analyze and compile data. These tools include TOAD, DataSpy, Erwin, and many other effective modeling and architecting data tools and software.

6) Ability to work with unstructured data

This is one area where a solid background in education helps a lot. As a data scientist, not every aspect of the work involves calculations, and even if it does, one needs to be able to determine which data is relevant in what equation for the outcome to make business sense. A high level of competence in critical thinking is therefore required to be able to maximize the benefits of massive data from social media, video feeds, and other sources.

7) Having intellectual curiosity

You surely can't afford to have someone as a data scientist if he or she relies solely on memory for remembering formulas and facts and other information found in books. A data scientist needs to be eager to learn more about things that are happening in the world, which can impact the cost of doing business, improve efficiency in doing business, or generally impact profitability. Intellectual curiosity is one of the important soft skills Frank Lo, DataJobs.com's founder, wrote about back in 2014 as a guest blogger for Burtch Works. It is a non-technical skill, but it makes a data scientist stand a cut above others in the field of analytics.

8) Competence in understanding a business situation

To help your organization become better, you can bank on someone with great business acumen rather than one who is least interested in matters of business. Business acumen is another one of those non-technical skills that are nevertheless very helpful when a data scientist is analyzing data to solve business problems. A data scientist with business acumen is able to analyze data and deduce what the lead problem for your business is and what factors are escalating it. This is very important because it then helps you in prioritizing business decisions in a manner that allows you to protect the business, even as you capture fresh business opportunities. Overall, your organization can better leverage whatever data it has.

9) Effective communication skills

Have you ever listened to educated people who address a group as if to impress them? That shouldn't happen with a data scientist. The marketing or sales managers, for example, couldn't care less if you used the IFFEROR or VLOOKUP formula, the INDEX+MATCH formulas, or even the popular IF formula in your data analysis. All they want is information that tells them the current market situation as deduced from available data and possible scenarios from that same mass of data. Data analysis is meant to provide quantified insights into the business situation, and unless those insights are communicated clearly and effectively to the relevant decision makers by the data scientist, all resources spent on data analysis will have been wasted.

Further Demystification of Data Science

Have you heard someone describe data science as the sexiest job of the 21st century? Well, those were the sentiments of expert contributors of the magazine Harvard Business Review, Thomas Davenport, and his counterpart, D.J. Patil. Although there are some skills that would make you better placed as a data scientist, different employers often have different needs in mind when they advertise for a data scientist. Thus, it is important that you do not give advertisements a casual look but a deep one, reading the details mentioned therein. In short, when it comes to daily practice, the term data scientist is a blanket term often covering four categories of engagements as follows:

1. Data Analyst

What we are saying here is that although you do not have the qualifications of a data scientist mentioned earlier, you would be doing yourself disservice if you are great at Excel and have some knowledge of MySQL database but then dismiss a job advertisement just because its banner says “data scientist.”

At times, people are hired to fill in the vacancy of a data scientist, yet they end up spending their days spooling reports and data from MySQL or doing work on Excel pivot tables. Other times, they are called upon to work with a team on some Google Analytics account. Although you may not be doing the deep work expected of a data scientist, the tasks just mentioned are not irrelevant or useless. Rather, they accord you a platform to practice the basics of a data scientist’s job if you are new in the field. In fact, if you were ever hired for such a post, you can take the opportunity to explore the additional use of data analytics beyond what the company demands, hence expanding your skill set.

2. Data Engineer

When they advertise for a data scientist, and somewhere in there, you find the mention of duties you are comfortable with as a data engineer, do not let the job slip by. If the job description includes things to do with statistics, machine learning expert, and such other roles that you can do as a software engineer, it is fine to go for it. Alternatively, you may find the job description being one of building data infrastructure for the organization. Companies often make such advertisements when they have too much data around, and they do not want to discard it because they feel they are likely to need it sometime in the future. Other times, they see the data as being important, but since it is unstructured, they do not know how to put it to good use.

3. Statistician

Sometimes, companies dealing with data-based services need someone who can handle consumer-focused data analysis or even someone who can deal with intense machine learning activity. Most of these companies actually run a data analysis platform. The tasks involved call for someone who is great at mathematics or statistics, or even someone with a background in physics – presumably people who are looking forward to advancing their academics along those lines. For this reason, if you see a data scientist's job posted by a company that produces data-driven products, do not hesitate to apply, especially if you specialize in mathematics, statistics, or something relating to calculations and analysis of figures. You could be just the person with the skills they require at that time.

4. General Analyst

You may find a big company seeking to employ a data scientist, but when you look at the details, you find that the tasks the person is to handle deal more with data visualization or even machine learning. It is alright to apply for the job because it means that the successful candidate is going to join an experienced team of data scientists, and the company just wants someone to help out in the lighter chores. What a great learning experience it would be for you, especially if you haven't had much experience in the position of a data scientist! In short, if you are skilled in some big data tools, such as Hive or Pig, you should be comfortable applying for this job irrespective of the heading, "data scientist."

5. Data Architects & Modelers

With the increasing amount of data that companies are required to keep and maintain, it is necessary to have people who can transform data from all the various systems into databases that can make more structured sense of it. This newly structured data is used to identify risk, fraud, and alert triggers. The Modelers and Architects work with development project teams to make sure that any system change gets sent down to the data repository and will be in a format that can be used for the reports and functions needed to use that data.

The Future Role of the Data Scientist

Experts are debating on the future role of data scientists. Will they become extinct by 2025? In the different regions of the data world, there is an emerging opinion that data scientists could become unnecessary personnel in anywhere from 5 to 10 years. Others say this could become the reality in 50 years.

There are those who say that the need for a data scientist will always exist. The core group who say that data scientists will never become extinct based this statement on what is called "expert-level tasks." The basic assumption is that there are some data science tasks that are too complicated for robots or automation to perform. The need for human creativity and innovation in data expert-level tasks will always be needed. Robots won't be able to think outside the box when data interpretation calls for new data model methods to be applied or built to solve unknown or hidden business issues as they arise.

The other side says that all data expert- level tasks, regardless of level of complexity, will be automated within 5 to 10 years. Software tools will be able to complete complex tasks that data scientists now perform. One example that was given is that with software tools like Tableau, the very difficult task of visualization has been mastered through an application. Second-generation data science companies are making software tools that will improve company workflow and automate data interpretation. Currently, this issue is far from being solved, so only time will tell whether the data scientist's job is in jeopardy or not.

Chapter 18: Big Data Impact Envisaged by 2020

Have you realized that it is not just advertising and marketing that organizations are taking online these days? If you consciously think about what takes you online on a daily basis, you will realize that a good part of it is business, often because you feel you can find a wider choice of items there, plenty of information about the product or service that you want, and even a wide range of pricing. The web also seems to provide much more entertainment than you could physically reach in a short time and, most probably, at a relatively smaller price. When it comes to connecting with others, individuals, private and public institutions, and all manner of organizations are taking their communication online, where they can reach a wider audience much faster and more cheaply.

How Does Moving Online Impact Data?

First, it means that the amount of data being generated on a daily basis is growing exponentially, and we can no longer ignore big data. Second, even before anyone can speak of the ability to analyze and organize the massive data, tracking is already a mammoth challenge in itself. Would you believe that internet users are generating data in excess of 2½ quintillion bytes each day? This includes automated feedback, such as traffic monitors, weather-related trackers, and all manner of transactions. Is this a good thing?

Well, potentially, it is. However, there is the question of what sources are reliable and which ones are not, what data is relevant to your scenario and which one is not, and so on. As you may already know, having massive data before you can also be overwhelming. That is why big data has created room for a unique business, where you get people with special training to make good use of big data. That potential power of big data is what specialized companies seek to help you unravel and take advantage of by handling big data that affects your organization.

In fact, specialized technology has started to make big data management convenient. A good example of this is Apache™ Hadoop®, an advanced database management technology that takes you beyond the consolidation of information towards the improved efficiency of your business and increased profits. As long as organizations are open to the use of advanced technology that makes big data analysis and management convenient, the world is headed for better times.

Instead of being overwhelmed by high data traffic and an accumulated mass of data, organizations are going to unleash the power of such data, and if they are in education, they are going to drastically bring down the cost of education. If they are in meteorology, they are going to have better comprehension of certain complex phenomena, such as the weather. Those in business will be able to raise their productivity by drastically cutting on all manner of redundancies, and the job market is going to have better correlated data sets that will help to match job seekers with potential employers based on the skills offered and needed, respectively. Some of these things are already happening, and they can only become better.

The story of big data does not end with the reduced cost of doing business and acquiring education, increased efficiency, and productivity, as well as increased profits. Ongoing research points to the potential of improving the fight against crime, significant improvement in web security, and the ability to foretell the likelihood of an economic or natural disaster sometime in the future. In fact, as long as there is ongoing work regarding big data and its intricacies, the world is likely to witness bigger and more radical changes, much of it for the better.

What's the Market Like for Big Data?

As previously explained, there are many innovators trying to create analytical and data management tools that will turn massive data into an economic advantage. In 2012, the big data market was worth $5 billion. If you remember what we said earlier that big data has been growing exponentially, you will not be surprised to know that if the trend today continues, the market for big data is projected to reach $50 billion by 2017.

Specific Ways Big Data Can Significantly Impact Life – Soon

1. Websites and applications functioning better

In this regard, it is expected that it will be much easier and more convenient to navigate websites and make sense of data, but and these sites are believed to become much safer than they are today.

This is particularly so because big data can be useful in identifying and tracking fraudulent activity on a real-time basis. Big data can introduce fresh and clear visibility on an organization's website while making it possible to foretell when attacks are imminent. Innovators have actually already begun designing programs geared towards safeguarding data against destructive intrusion. For example, there is the machine learning program known as MLSec that you can find at MLSec.org. It uses algorithms under supervision to locate networks harboring malicious programs. It must be very encouraging to web users to learn that this machine learning program has been proven to be accurate at a rate of 92% to 95% for every case tested.

How Bad is the Security Situation Today?

Based on 2012 statistics:

- Of all the websites hacked in this year, 63% of the owners did not realize they had been invaded.
- In fact, 90% of web owners did not even seem to notice anything strange at all going on within the site.
- Notably, 50% of the web owners learned that their websites had been hacked from their browser warnings or even warnings from the search engine they were using.

1. Higher education becoming more accessible

Why is it so exciting that cost of education would fall? Well, accessing higher education is generally a problem for the average person in most countries. In the U.S., which is considered by many as the land of plenty, the cost of tuition rises at a rate that is double that of healthcare. When you compare the hike in the cost of education to that of the country's Consumer Price Index, it reaches four times as high.

Luckily, just as websites are becoming better designed and protected, something is happening to make education more easily accessible to a larger number of people. A good example is the emergence of online sites offering great courses. The Khan Academy found at khanacademy.org, Big Data University found at BigDataUniversity.com, Venture Labs found at venture-labs.org, and Coursera found at courser.org are just some examples of institutions that are offering higher education online at a much lower cost – and sometimes even free of charge – than conventional tertiary institutions.

The good thing about many of the courses offered in this manner is that students are tested on how well they have clinched the skills taught, particularly because these skills are applicable to the current high-tech environment. For example, Big Data University teaches Hadoop, along with some other big data-related technologies.

2. Relative ease in getting a job

Have you ever given thought to the number of job seekers there are worldwide? It must be a staggering figure. In fact, on a monthly basis, the number of job searches on the web alone has hit 1.5 billion. On the positive side, there are websites that have already begun to match job seekers with potential employers by amassing valuable information on various employers, as well as information on interested job seekers. One such website is indeed.com.

3. Improved road safety

Are you aware that car accidents are the main cause of death in the category of youth between 16 and 19 years of age in America? While it is dangerous to drive while under the influence of either drugs or alcohol, 75% of the deaths in this category are not related to alcohol or drugs. What this means is that there are many accidents that occur merely because of poor judgment. It is envisaged that as scientists continue to work on advancing the technology being used on computers, big data is going to help them predict the behavior of drivers on the road, as well as the move a vehicle is about to make at a certain point.

This route is geared towards reaching a point where cars on the road can exchange data so that drivers in different cars can see up to three cars that are ahead of them, three that are immediately following them, and three on either side of them at any one time. In fact, it is said that big data will enable the drivers in data swapping vehicles to see the posture and focus of a driver in a car near theirs. This may sound a little far-fetched, but think about it: Haven't Google's self-drive cars taken the auto industry to a totally new level?

4. Ability to predict business future

What this means is that organizations will be able to use software like Hadoop to analyze the data at their disposal in a way that will bring future possibilities to the fore. The speed and accuracy with which data will processed will enable organizations to take prompt action where there are business opportunities emerging, where damage control is needed, and where other actions that call for accurate assessment are required. There are already big organizations using Hadoop with great success. These organizations include eBay, Twitter, FaceBook, Disney, and others. The demand for Hadoop is rising rapidly. IDC, a renowned market research firm, has predicted that by 2016, the conservative worth of this software will be $813 million.

Another good example is Recorded Future, a technology company based on the web. This organization provides data analysts with security intelligence that they use to keep their information safe. It puts businesses in a situation where they can anticipate risks and also capitalize on business opportunities by unlocking predictive signals using clever algorithms. There are other examples already helping businesses prepare for eventualities, but suffice it to say that as technological development continues, it will become all the more possible to leverage data, hence avoiding surprises.

5. Ability to predict matters of weather

The ability to predict the weather brings the advantage of being in a position to protect the environment better. Such protection, in turn, brings in huge savings to the country and the world at large, considering the massive expenses that are brought about by weather-related disasters. For example, weather- and climate-related disasters cost the U.S. losses in staggering figures that are actually in excess of $1 billion. Here, we are talking about such disasters as drought, wild fires, incidences of flooding and storms, and other unfortunate events.

Things are already looking bright on the data technology front. For instance, there is the Joint Polar Satellite System (JPSS), which is set to be launched in 2018. The JPSS, which will have the capability to utilize sensor technology, will also use data to determine a hurricane's path or a storm's path well before these disastrous events occur. This will then give everyone concerned time to plan what to do to safeguard life and property. As CNBC News has already noted, situations that relied on guesswork some years back are slowly but surely becoming something to predict with precision through predictive science.

6. Healthcare becoming all the more efficient

Big data is expected to bring improvement to the health sector, not just by raising efficiency levels in service delivery but also by customizing services to suit respective consumers. McKinsey & Company, an advisor to the management of many renowned businesses, says that between 50% and 70% of business innovations depend to a great extent on the capacity to capture customer's data and not as much on external analytics. McKinsey & Company relies heavily on qualitative and quantitative analysis to be able to give helpful advice to management.

It is said that 80% of data in the medical sector is unstructured. However, with the trend the health sector in the U.S. has taken in using big data creatively, great improvement in service delivery is anticipated. It is actually envisaged that big data is going to help the sector increase value through efficiency, adding value in excess of $300 billion each year. By the same token, expenditure is anticipated to be reduced by a good 8%.

In treating patients effectively, caregivers benefit a lot from available patient data. This is because such data helps caregivers provide evidence-based advice. Currently, a medical center within Boston called Beth Israel Deaconess is putting a smartphone app in the market as a means to help medical caregivers access 200 million data points. These data points are expected to make data concerning around 2 million patients available. There is also Rise Health, which utilizes the accessible mass of patient data by analyzing it from all dimensions and aligning it with the health providers' goals to improve healthcare through fresh insights. Overall, big data brings speed to innovation. A fitting example is the project on the human genome that took 13 years to complete; today, it would only take a couple of hours to accomplish it.

Chapter 19: Benefits of Data Science in the Field of Finance

As far as data is concerned, there is no longer a shortage. There may even be an excess of data if you consider the traffic going through social media, real-time market feeds, transaction details, and elsewhere. The volume of data available for use in the finance sector is almost explosive. Its variety has also expanded, and even the velocity at which the data becomes accessible has sharply risen. This scenario can either take organizations to heights unknown before or leave them dumbfounded from the feeling of overwhelm caused by the data influx.

Given that organizations are in business to succeed, they have learned that the best way to utilize this flood of data is to engage data scientists. A data scientist is a guru who takes the data, explores it from all possible angles, and makes inferences that ultimately help him or her make very informed discoveries.

A Data Scientist utilizing Data:

- Identifies and captures fresh data sources, analyzes them, and then builds predictive models. The data scientist also runs live simulations of various market events. All these make it possible to visualize the reality of possible situations even before any measures have been implemented. Data science helps the organization to foresee trouble well in advance and prepare for it accordingly, as well as foretell future opportunities as different factors play out in the business arena and the environment in general.

- Utilizes software like Hadoop, NoSQL, and even Storm to optimize data sets of a non-traditional nature, like geo-location and things like sentiment data. After that, the data scientist integrates the data sets with that which is more traditional, like trade data.
- Takes the precautionary move of ensuring there is ample raw data in storage for future reference and analysis. In that regard, the data scientist finds the relevant data in its raw form and selects the safest and most cost-effective way of storing it.

The expertise of data scientists in utilizing big data is being made even more convenient by the emergence of other technology-based storage facilities. There is, for example, cloud-based data storage, as well as analytical tools that are not only sophisticated in the things they can accomplish but also cost effective. Some are tools that you can access online free of charge, presented as open-source tools. In short, there is a whole range of financial tools that are at the disposal of data scientists, and they are being put to use to transform the way of doing business.

A Data Scientist Understands People's Sentiments

Is it strange that a data scientist should be able to analyze people's sentiments simply from data? Well, there are suitable tools for that. In conducting sentiment analysis or what you can aptly call opinion mining, a data scientist makes use of natural language processing, text analysis, and computational linguistics to consolidate the relevant material required for the process. There are already firms using sentiment analysis to improve business. Some good examples include MarketPsy Capital, MarketPsych Data, and Think Big Analytics.

How Firms Work through Sentiment Analysis:

- By building algorithms surrounding sentiment data from the market, for example, through the use of Twitter feeds. These feeds give ample data when incidences with big impact, such as a terrorist attack or even a big storm, occur.
- By tracking trends, monitoring new products introduced to the market, responding to issues affecting your brand, and in general, improving brand perception.
- By analyzing voice recordings right from call centers when those recordings are unstructured, then recommending ways of reducing customer churn or, in other words, recommending ways of raising customer retention.

Considering that most of today's businesses are customer-focused, the need for a service analyzing data in order to have a vivid picture of what customers feel about a brand cannot be overstated. In fact, data companies have emerged to fulfill this need, and their role is that of intermediaries which gather data, identify sentiment indicators, and then sell that crucial information to retail businesses.

Data Scientists in Risk Credit Management

The way in which data scientists utilize the amount, frequency, and variety of data available online has enabled firms to offer online credit with minimal risk. In some places, potential investors simply fail to access credit because there is no way of giving them a credit rating. Nevertheless, a lender or financier needs to know the extent of risk involved any time lending is about to take place. Luckily, with big data and expertise from data scientists, internet finance companies have emerged, and they have found ways of approving loans and managing risk. Alibaba Aliloan is a good example of online lending that is enabled by the use of big data.

Aliloan is not a conventional bank but an automated online system that offers flexible small-sized loans to online entrepreneurs. The recipients are often creative and innovative persons who find it difficult to get credit from traditional lenders, like banks, simply because they have no collateral.

Big Data Reducing Risk of Online Lending

Let us use Aliloan as an example of online lending:

i Alibaba monitors its e-commerce platforms, as well as the ones it uses for payments, to understand customer behavior and financial strength. After analyzing the customer's transaction records and customer ratings, as well as analyzing related shipping records and other related information, Alibaba is able to determine the loan ceiling to set for the customer while considering the level of risk learned after the comprehensive data analysis.

ii Alibaba also gets the online findings confirmed by third-oparty verifiers, even as it seeks other external data sets to cross check the online findings against. Such helpful external data sets include customs and other tax records, electricity records, and other utility bills.

iii After granting the loan, Alibaba keeps tabs on the customer's activities, monitoring how the customer is utilizing the funds provided for investment. The lender generally monitors the customer's business strategic development.

Other companies that are offering loan facilities by relying heavily on data scientists' expertise on big data are Kreditech and Lenddo, both of which offer small loans on an automated basis. They have come up with credit scoring techniques that are very innovative yet very helpful in determining a customer's creditworthiness. There are also cases where much of the data used to assess a customer's position is from online social networks.

Real-Time Analytics Improving the Finance Industry

Any decision maker in the finance industry will tell you it's not enough to have data somewhere within reach; it matters when you analyze it. As many people dealing in critical thinking will confirm, it is not possible to make an informed decision before you analyze the data. Thus, the longer you delay the process of data analysis, the more you risk business opportunities passing you by, and the higher the chance of other things going wrong in your business. However, with the skills that data scientists have relating to big data, time lags are no longer a handicap in the finance sector.

How real-time analytics helps businesses:

1) Fighting fraud

Today, it is possible to detect attempts at fraud through data analytics relating to people's accounts. Institutions like banks, credit card companies, and others have gotten into the trend of remaining on top of things as far as fundamental account details are concerned, courtesy of big data. They want to ensure that they know whether your employment details are up to date and your physical location, too. They analyze and see the trend of your account balances, learn about your spending patterns, analyze your credit history, and have such other important information at their fingertips. Given that data is analyzed on a real-time basis, data scientists ensure that there is a red flag triggered whenever there is suspicious activity taking place or even when an attempt is detected. What happens then is that the account gets suspended so that the suspected fraudster cannot continue operating, and the owner receives an alert to that effect instantly.

2) Improving credit ratings

Can anyone give a credit rating without ample data available? Needless to say, the rating is only credible if it factors in current data and not just historical data. That's why big data is so important in this era in which credit ratings play a crucial role in determining the level of risk you carry and the amount of credit you can be allowed by lending institutions to enjoy. The fact that data analytics takes place on a real-time basis means that customers' credit ratings are up-to-date, and they provide a reasonable picture of the customer's financial capacity. In any case, most of the categories of data necessary are already covered online, including assets in the customer's name, various business operations the customer is engaged in, as well as relevant transaction history.

3) Providing reasonably accurate pricing

This pricing factor cuts across products and services. In financing, it may mean that a customer can get a better rate of interest levied on money borrowed if the current rating is better than before. For insurance, a policy holder can enjoy benefits derived from data analysis, issuing timely warnings over accidents ahead, traffic jams that may affect the driver, weather conditions, and such other information that can help in reducing the rate of accidents. With a policy holder having a clean driving record or at least an improved one, it is possible to win a discount on the price of an insurance policy. It also means that insurance companies will have less to pay out as compensation.

Overall, the cost of business goes down for everyone whenever data analytics is used. Today, that is primarily because of the benefits accruing from real-time analytics. In fact, major financial institutions like banks have made PriceStats the order of the day. This online firm, which began as a collector of daily inflation rates for a few countries in South America, now monitors prices for around 22 countries (as at 2015), with the aim of providing daily inflation rates for those economies. This means you can easily follow the fluctuating or steady trend of inflation rates and tailor your business actions accordingly. PriceStats also has a lot of useful data based information for the U.S., which is the largest world economy.

Big Data Also Great for the Customer

For those who may not be aware, many institutions, including banks and a good number of other financial institutions, do pay to acquire data from a myriad of retailers and service providers. This underlines the importance of data, particularly when you have the capacity to analyze it and use it where it matters most. In fact, all data is important depending on the reasons someone wants it for. You do not want to store data just for the sake of keeping it. Data held within your system with nobody accessing it is an unnecessary distraction. Even when data comes your way without you requisitioning for it, you can easily get overwhelmed unless you have a good processing plan in place.

That is why it is important to have a data strategy that is inter-departmental so that you can identify the category of data to provide to another department instead of discarding it. You can also determine which portions of data to get rid of straightaway. The reason this chapter places emphasis on the contribution of an analyst to data handling and utilization is that not everyone can make good use of data from scratch.

However, a data scientist has the necessary skills to handle big data from A to Z.

This is why big data is helpful, particularly when there is customer segmentation.

In the case of institutions that pay to receive data, their aim is to use the data to create a 360° visual image of their customer. As such, when they speak of KYC (Know Your Customer), they are speaking from a point of credible information, and that reduces the risk of doing business with these individuals. Can you see predictive analytics coming into play right there? This aspect of using big data to have an overall understanding of the customer has been emphasized by Sushil Pramanick, a leading figure with IBM analytics. Pramanick also happens to be the founder of The Big data Institute (TBDI).

Improving Business through Customer Segmentation

Once you can put together customers who have the same needs and probably the same financial capacity, customers who have similar consumer tastes and are in the same income bracket, customers who are in the same age bracket and are from a similar cultural background, and others who match in various ways, it becomes relatively easy to meet their needs.

a) You can conveniently design customized products and services for them as a target group.

b) You can adjust the manner of relating to them with the aim of retaining them as customers, this avoiding customer churn.
c) You can tailor your advertising and marketing approaches to appeal to these target groups.
d) You can re-engineer products or develop new ones with specific groups in mind.

Chapter 20: How Data Science Benefits Retail

As previously mentioned, data in circulation is increasing in volume each year at a rate that can only be described as exponential, and its variety and velocity is also shooting up. Retailers who are smart in business know there is value in data, but they may not be certain about how to collect and analyze data at every possible interaction for use – where use, in this case, implies more business profitability.

This is one reason why the demand for data scientists has risen in recent years. Experts like McKinsey have weighed in on the benefits of employing data analytics in the retail business. In a report they released in 2011, McKinsey projected that the retailers going with big data analytics are likely to have their operating margins rise by a good 60%. That was a way of reassuring retailers that data scientists have the capacity to turn big data upside down and inside out – whether it is structured or not, internal or external, – by organizing it in a way that makes business sense, ultimately helping the retailer create gold out of the mound of apparent clutter.

Retailer's Way of Reducing Cost and Increasing Revenue

- Receiving recommendations that are customized to the needs of the individual retailer.

- Taking advantage of sentiment analysis on data from social media or call centers or even the ones displayed through product reviews. All such data is important in giving a clear picture through customer feedback and gives depth to market insights.
- Improving customer experience through predictive analytics, with improvement being both online and offline.
- Improving how retailers lay out their promotional displays and such other merchandizing resources by utilizing heat sensors and even image analysis to gain a better understanding of customer behavior patterns.
- Utilizing video data analysis to identify customer shopping trends. This analysis is also helpful in identifying cross-selling opportunities.
- Making consistent daily profits because of the ability to respond to the information derived from internal and external data. Would there be any benefit, for instance, in taking advantage of the reduced cost of raw materials if the weather did not allow for deliveries of the final product? Thus, data analysis puts the retailers in good stead because they can plan for business with prior knowledge of the weather, economic forecast, traffic reports, and even about whether it is low or high holiday season, among other details.
- Growing revenues at a faster rate simply because it is possible to perform a detailed market analysis
- Making use of product sensors to communicate, on a real-time basis, important information concerning post-purchase use.
- In term of marketing, retailers will be cutting cost by offering personalized offers via mobile devices. They

are also able to deliver location-based offers, which are much cheaper than generalized marketing.

- Retailers are communicating real-time pricing, making it possible to issue prices based on metrics on a second-by-second basis. In this manner, retailers can also access their competitor's data regarding consumer behavior and other factors relevant to business.
- Identifying appropriate marketing channels based on the indicators from analytics on segment consumers. This helps the retailers to optimize their Return on Investment (ROI).
- Making use of web analytics and online behavioral analysis to tailor marketing offers accordingly.
- Retailers are benefitting in the area of supply chain logistics. In this regard, data science helps when data scientists use data to track inventory and perform general inventory management on a real-time basis.
- Retailers are assuming a position in which to optimize their routes of product delivery through the telematics of GPS-enabled big data. Of course, with well-analyzed data, retailers are bound to select the cheapest, safest, and fastest route.
- The area of supply chain logistics is benefitting from both structured and unstructured data, as retailers are able to foretell likely customer demand well in advance.
- Retailers are getting the opportunity to negotiate with suppliers once they read the scenario from the analysis of available data.

Chapter 21: Using Big Data in Marketing

Marketing is one of the areas of human life that can most significantly benefit from the use of big data. Marketers and data analysts both have the irresistible need to know more about those they are looking into. Thus, big data analysis can add huge amounts of value for marketing experts when done correctly.

Nearly every human action is quantifiable. Everything you do, say, or see can be turned into numbers, put into tables, and crunched for trends. The implications of big data analysis for marketing are as wide as you can imagine. Everything from how much people are spending to what kind of merchandise they prefer can be determined through big data analysis.

A decade or two ago, it would have been nearly impossible to describe customers in such detailed ways as possible today. Back then, the best we could do is track how people responded to certain marketing methods, such as how many subscribers we got based on an email campaign. Meanwhile, we can have so much more information today. Things like clicking behavior patterns have allowed marketers to place ads in a better way, for the purpose of increasing the number of clicks on these ads and, consequently, the revenue.

Google Trends

Using Google Trends has become one of the most widespread and useful ways of integrating big data into your marketing strategy. Google Trends enables even small business owners to get the most out of big data without having to invest anything into the analysis. Google does all the work, and they reap the rewards. This super useful tool tracks what people are searching for on Google and is probably the most reliable way to discover what is trending in the world in general and in any particular industry. This is an immensely useful tool for marketers, who can now know exactly what people are interested in, catch trends early, and use this information to bomb their customers with exactly what they are looking for. Designing marketing campaigns and coming up with new product ideas have never been easier, thanks to Google Trends. The best part about it is you do not need to crunch the numbers yourself.

Defining the Ideal Customer Profile

In the past, marketers were limited to knowing only general information about the public and had to define their customers by trial and error. Now, using big data analysis, it is possible for a business to acquire incredibly comprehensive information about the general public and particular parts of the population, as well as better define the ideal customers. While marketer's intuition can still make the process easier, it is now possible to know exactly what age, gender, and interests your target buyers share. Knowing this kind of information makes it possible to make better and more targeted campaigns aimed directly at these specific buyers. Certain studies have shown that adding big data analysis to the customer profile defining process improved the results of marketing campaigns by up to 30%.

Real-Time Personalization

In order for marketing to be successful, it must be done in a timely fashion. Both timely delivery and quality of the content will be needed in order to market products or services successfully, and both these things can be achieved through the proper use of big data. Big data has become so comprehensive that it is now possible to track people's tendencies in real time and know exactly what your customers want and when they want it. Knowing this enables marketers to send out highly successful marketing campaigns, whether it is via emails or other methods, and fill these campaigns up with useful and relevant content for their target audience.

Identifying the Right Content

Before big data became a hit, we could never be really sure as to whether a piece of content or a type of content was working or not. We would rely on hunches and expectations and do our best to provide content we believe may send the customers down the sales tubes and convert them to a sale. Today, we can use big data analysis to determine exactly if and how much a certain campaign worked and go as deep as determining if a certain tweet converted well or not, if a Facebook post got us new customers or not, how much they spent, and what they bought. Big data helps us understand what worked and what did not, and it can also help us anticipate future content that customers may want to see. Designing content based on this information will have an incredible impact on final sales, and we are able to skip the guesswork by using these strategies.

Predictive Analysis for Lead Scoring

Predictive analysis is used to predict the future behavior of customers and, in the case of lead scoring, it is one of the most useful strategies marketers can employ. By using the data a company has collected in their CRM, as well as externally collected data such as those of Google Trends and other internet services, a company can better determine their lead scoring strategy. Getting new leads can be a very costly process but using big data analysis has shown a reduction in lead price of up to 50% for many companies across various industries.

Customer Geolocation

The geolocation of your customers used to be a difficult thing to determine. With the availability of big data, we are now able to determine with precision where our leads and visits are coming from, and this allows us to personalize the content we are serving to perfection. For instance, let's assume we are using blog posts to drive sales. Now, using big data, we can determine where the most readers came from. There may be factors we cannot know that are driving Colombian traffic (for instance) to our blog posts. In this case, it may be a good idea to create posts in the Spanish language to better connect with the audience from Colombia that is already reading your blog, even though it is in the English language. This is one of the ways we can use geolocation data to better connect with our customers.

Evaluating the Lifetime Value

While this may be a slightly bad way to look at customers, the reality is that for a marketer, every customer is a number – a value. With the use of big data analysis, marketers can determine the exact value an average customer will be worth over their lifetime – over a year or over a month.

There are various pieces of information that you will gather into your CRM in order to determine how valuable your average customer is. For starters, the average amount of money a customer spends on your products will be a very important number, of course. Beyond this, you can calculate the average costs of things, such as support time an average customer needs, and come up with a final number that is the estimated value of a customer.

Looking at the Big Picture

Using big data for marketing purposes can be a double-edged sword. Using certain metrics can sometimes increase one part of your business but decrease another. I will use an example to demonstrate what I mean. Let us say we have sent out an email campaign, and the result was a big increase in the number of sales made. Looking at sales made per emails sent, we have made a significant profit, and we celebrate. However, if we are not looking at the big picture and examining all the data available, we may miss a decrease in a different metric. For instance, this email campaign may have caused many people to unsubscribe from our list. If this happens, we may be losing future sales, and continuing to push this campaign may end up sending customers away instead of bringing new ones in and, in the long run, the campaign may be a complete failure.

Comparing Yourself against the Competition

Modern data analytics has allowed businesses to compare themselves against other companies in the industry and see how they stand. In the past, it was always guesswork as no one would let you have a look at their private information, such as sales numbers or website visits, and the information was simply not available.

Today, you can easily see your own metrics, and by using such tools as social media analytics tools, you can gain access to important data regarding your competition. Is our competitor getting more retweets? Do people prefer his Facebook posts to ours? This is the kind of data we can find and use to gain a competitive edge against other companies.

Search engine data can also be used and compared to your own information to determine how well a competitor is doing. There are various tools out there that allow us to see how a website is ranking for certain keywords and, in general, this type of information will clearly let you know who the leaders are in your industry and how you are doing in comparison to them. In the modern world, things like number of website visits and social media activity can truly be used as a reliable statistic reflecting where a business stands.

Importance of Patience

Big data can certainly help any marketing team out there, but it will not do miracles overnight. The data collection process is one that will consume a great deal of time, and data analysis can be just as time consuming and costly. Despite using outside numbers, such as Google Trends, it will still take plenty of work and human labor to get to the point where you are using this data in the right way. More importantly, this information will not absolutely propel your business to the top but would rather help you maintain your business and gradually grow it if used right. It is very important to be patient, especially while collecting the information, and not jump to conclusions based on a small sample size of customers.

Chapter 22: Data Science Improving Travel

The world of travel has always had an appetite for lots of data. It has always encountered lots of data, too, even when stakeholders are not deliberately searching. The biggest challenge has mostly been how to store the data, as well as how to analyze it and optimize its use. The data that can be of benefit to the travel sector is massive, considering that all facets of life are involved, including people's cultures, air fares, security in different geographical areas, hotel classes and pricing, and the weather, among many others. For this reason, this sector cannot afford to ignore data science, being the discipline that will help in handling the large data sets involved in travel.

How the Travel Sector uses Data Science

There are many benefits that come with the use of big data, which is possible when you have a data scientist working with you. Some of the benefits that come with the use of big data in the travel sector include:

- Ability to track delivery of goods. This is possible, whether it is a freight shipment involved, a traveler on the road, or a voyage. Major online firms, such as Amazon and Etsy, benefit from such tracking, especially because it is not just the seller who is able to track the shipment but the customer, too. This is likely to give customers more confidence in the company and their mode of delivery, thus making them repeat customers, which is obviously good for business.

- Analysis done at each data point. This is one way of increasing business because it means that different stakeholders will be able to share information from well-analyzed data. This ensures that nobody receives data that is irrelevant or redundant.
- Improving access to travel booking records. This involves the use of mobile phones to make inquiries, bookings as well as payments and other simplification of transactions
- Easy access to customer profiles, itineraries, positive and negative feedback, as well as other data from internal sources, such as sensor data.
- Easy access to external data, such as reviews from social media, the weather, traffic reports, and so on.

In short, the travel sector is a great beneficiary of big data, and it merges the data from both its internal and external sources to find solutions to existing problems. The same data is analyzed in a way that helps stakeholders anticipate with relative precision the likelihood of future events. The sector is also able to cut drastically on the cost of operation because organizations can make better choices when they have massive data analyzed in a timely manner, which is what happens when data science is employed.

One reason there is a lot of interest in big data in the travel sector is that the sector is lucrative, and it is projected to become even more rewarding in terms of revenues. Global travel is projected to grow sharply, such that by 2022, its value will have reached 10% of the world's gross domestic product (GDP). Of course, the main players involved realize the value of big data and want to optimize its use so that they get the best business intelligence to work with. That makes them see big bucks ahead.

Big Data Enabling Personalization of Travel Offers

Some years back, travel considerations were given based on a general classification of customers. Those in the high-income bracket were sent recommendations for certain facilities, those who regularly travel with children were targeted for different offers, and so on. While that may have increased revenues slightly, its impact is still peanuts compared to today's use of big data.

The reason big data is very significant in the travel industry today is that it enables a travel company or agency to tailor offers for individual customers by relying on a 360° view of the customer. Thus, the potential of offering ultra-personalized packages or facilities to individuals is not far-fetched when big data is at play.

Some data sets that help produce a 360° view

- Data directed at reading behavior

A good example of this is the regularity with which a person goes online and the websites this person regularly visits.

- An individual's posts on social media

The analyst can establish whether the person writes posts about travel on social media, whether friends speak about travel with this person, or even if there are any travel reviews contributed by friends.

- Data from location tracking
- Data on itineraries on various occasions
- Historical data about a person's shopping pattern
- Data reflecting how the person uses mobile devices
- Data regarding image processing

This list is by no means exhaustive. In fact, any data or information relating to an individual's preferences on travel would make the list. As you will realize, different data sets suit different people, while others are common to all. For example, the data sets targeting a person's behavior pattern will mostly be useful for a potential travel customer or one who is new to the sector. However, when you speak of historical patterns, then you are targeting old customers.

Big Data Enhancing Safety

If you consider the information transmitted between coordinators of travel from the pilot to the control tower and vice-versa, from the driver to the fleet headquarters and vice-versa, from traffic headquarters to all travelers, from electronic media to travelers at large, and so on, you will easily acknowledge that big data is a real life saver in the travel sector. In fact, vehicles and planes today are fitted with different types of sensors that detect, capture, and relay certain information on a real-time basis. The information relayed varies greatly and may include airmanship, the behavior of the driver, the mechanical state of the vehicle or plane, the weather, and so on.

This is where a data scientist comes in to design complex algorithms that enable the travel institution to foretell when a problem is imminent. Better still, the work of the data scientist helps to prevent the problem before it crops up.

Here are some of the common problems that big data helps address:

- If there is a part of the vehicle or plane that is not working well, it is replaced before the damage becomes too serious for the part to be salvaged. Replacement or repair is also done to prevent an accident.
- In case the driver is missing one step or more when traveling, you need to pull him or her out of active duty and take him or her through some further training. This is necessary in order to maintain high standards of performance.
- For flights, big data helps to identify problems that can be addressed when the plane is in midair. For those that are found to be difficult to address when the plane is in the air, the company places a maintenance team on standby so that they can rush to check the plane immediately upon arrival.

From the facts provided, it is safe to say that big data plays a crucial role in averting accidents during travel, whether by air, sea, or even by land. This can be easily appreciated when you think about the transmission of weather-related data and information on a real-time basis.

Up-Selling and Cross-Selling

Sometimes, big data is used in up-selling and other times, in cross-selling. Just to recap, vendors are up-selling when they are trying to woo you into buying something that is pricier than what you were seeking in the first place, which is a great marketing attempt. As for cross-selling, vendors just try to get you to buy something different in addition to what you were searching for initially.

In case you want to travel by air, you are likely to receive many other offers together with the one you inquired about in the first instance.

Here are examples of up-selling and cross-selling:

- You may find an offer in your inbox that is personalized and is a form of cross-selling.
- You may find yourself being booked for Economy Plus, which means you are being up-sold. With the advantages of additional leg room and an opportunity to recline further on your seat, it is tempting to take up the offer even if it means forking over an extra amount of money.
- You are likely to find discounted offers of hotels partnering with the airline you are using for travel.

- You could also receive an offer for dining, courtesy of the steward, and this is on a complimentary basis so you get a coupon.
- You may find yourself looking at an ad during an in-flight entertainment session, trying to attract you into using a certain city tour on arrival or during your local travel.

Companies, particularly those in the tourism sector, are partnering and using big data to cross-sell with each other. The usual suspects include airlines, hotels, tour van companies, and those offering taxi services.

Chapter 23: Big Data and Agriculture – Using Big Data to Feed People

Agriculture is one of those fields of human life that is always associated with primitive and old-fashioned things. Putting agriculture and data in the same sentence is something most of us never think about, but agriculture is actually another field where big data can be used to greatly help those involved.

Agriculture is the biggest industry in the world, and to date, it is still one that feeds the majority of people on the planet. Whereas it used to be just farmers plowing their fields, technology is now used in many ways to aid in the agricultural process. Things like agricultural machines have been used in the industry for decades, but things like data analysis for improving the efficiency of agricultural labor remain a fairly new idea.

Today, we are able to collect more information regarding agriculture than we ever could have imagined. Such statistics as soil elevation or final yield at harvest time can be gathered and analyzed to come up with a number of useful metrics that can be used to predict future events.

An average farmer cannot benefit much from raw data itself. While many of these things were known to farmers in the past, they had absolutely no way of using it. Today, agricultural companies hire teams of data analysts who are able to help the company and farmers make educated decisions, which can increase the yields of the fields and final revenue.

Increasing $ per Acre

No matter what business we are talking about, there is one factor that everyone is interested in, and this factor is revenue. We are all in it for the money, one way or another, and farmers are no different. They, too, have to make a living on a daily basis, and big data analysis is something that can greatly help with this.

A tool known as Climate Pro, for instance, offers farmers various pieces of information that they can use to increase their own revenue per acre of land. Nitrogen Advisor, for example, which is a part of Climate Pro, allows farmers to track the total nitrogen in their soil and gives them tips on how they should proceed in this regard. The owners of Climate Pro claim that using this tool alone could increase revenue by $100 per acre of workable land.

Another tool for farmers known as FieldScripts allows them to increase their corn yield. This tool breaks up your entire field into small sections and advises you on how much corn to plant in each part of the field based on what the soil is like. By using this tool, you can now plant the optimal amounts of corn in a field without doing any guesswork. All your data will be factual and based on big data. The tool can be connected directly to your tractor to changes the amount of seeds planted per acre automatically. This kind of technology use in agriculture is revolutionary and something that would have been unimaginable in the relatively recent past.

Feeding the Millions

Big data is often used for purposes that could be morally questioned and that only serve to help improve the bottom line, but what if we use big data to help with the agriculture of countries with populations that are starving?

India, for instance, has millions of acres of land to be worked on, and for the most part, the land is being worked by individual farmers who have no way of using or collecting big data. Companies, such as CropIn, believe that at the current rate, we will be completely unable to feed the world's population by 2050, but by using big data analysis, they are finding new ways of optimizing the agricultural process. Indian farmers could, for instance, benefit greatly if the entire country was being monitored and tracked in real time, with information being fed directly to the farmers and educating them on how to proceed with their process.

What about Trust?

This is the major issue that may be slowing down the progress of big data use in agriculture. Unlike many other businesses, farmers put their entire business at stake every year. If the yield is low or the crops are destroyed by a natural disaster, the farmer's family is in great jeopardy. This is why farmers rarely fully trust anyone, as it is simply very difficult for anyone to be trustworthy enough. This is why trusting big data analysis is even more difficult, as it may seem like an abstract concept to begin with.

Companies that provide much of this big data analysis, such as Monsanto, have been demonized and hated by much of the world for various scandals they were a part of and for potentially poisoning the very food we eat. Farmers are even less inclined to trust anything these people tell them.

For these reasons, it will take some time for big data to catch on fully in agriculture, and companies like Monsanto and other giants will need to earn the trust of lone farmers out there who hate putting their fate into anyone's hands, let alone those they consider to be destroying their industry. One thing is sure, though, sooner or later, every farmer in the world will be using big data analysis to optimize farming, as it is the prudent thing to do. This will become inevitable.

Will Big Data Save Colombian Rice Fields?

Big data can surely be used to help farmers across the globe. A particular example of where big data could be used to help improve the agriculture of an entire nation is Colombia, the rice fields of which have been experiencing inexplicable drops in yields for years.

The yield of these fields was gradually increased over a decade, but in 2007, the inexplicable decline began and could not be stopped. The reason for the decline was unknown, and no single person was able to really put his or her finger on the cause. While climate change was dubbed the lead suspect in this case, there simply was no evidence.

After some concerns about sharing the data on rice yields in specific parts of the land, government agencies began their surveys and soon had a reasonable amount of big data lined up. Crunching this data led to a great number of very specific inferences which that could potentially save the rice fields in Colombia.

The interesting thing is that particular regions of the county had shown different results. The city of Saldana, for instance, showed that it was the lack of daily sunlight that may be causing lower yields, while the rice cultivated around the city of Espinal was apparently not adapted to the increasingly warm nights in Colombia. The resulting inferences were that farmers of Saldana should likely move their planting dates to line up with the sunniest part of the year, while those in Espinal were likely to benefit from using a variety of rice that would be less harmed by the warm night and more in line with the local climate.

This type of descriptive and predictive data analysis could quite literally revolutionize the entire industry. Imagine applying this type of analysis to every single crop in every single part of the world. We could optimize the yields of all plants on the planet and never waste a field or a seed again. Obviously, using big data analysis is likely to feed the world in the centuries to come.

Scaling Up

With so many successful smaller projects around the world related to the use of big data to help farmers, it is about time that big data be applied to agriculture on a larger scale. It is likely that in the coming years, we will see more and more companies analyzing agricultural data and more farmers beginning to apply the revolutionary predictions made by these analysts to increase their yields and their profits successfully.

Chapter 24: Big Data and Law Enforcement

Big data is making critical crossroads in criminal investigations. Today, some law enforcement agencies aren't using data tools to their fullest potential. They will work independently of one another and maintain individual standalone systems. This is impractical and not a cost-effective way of running these agencies. Law enforcement is expected to do more with less. Manpower has been cut back, but crime still increases in some areas.

It is of benefit to these agencies to pool their shrinking resources into a data networking system where they network data across the board to help solve crimes and protect the public more efficiently.

Big data is the answer to finding effective solutions to stopping crime in a time when agencies are being asked to perform more tasks with less help. This is where big data comes in. They are being forced to streamline operations and still produce the desired results. Data has to be interpreted holistically to be able to streamline their operations. Data platforms make these objectives a reality.

The data platform can be placed in the human resources system or even in the plate identification system. It doesn't matter where the platform is integrated; it just has to be inserted into one of the data systems so that there will be a foundation from which criminal activity can be dissected, analyzed, and translated into effective crime-fighting techniques. The following are the challenges facing law enforcement:

- Law enforcement is being straddled continually with shrinking financial resources.
- Limited money equates to less manpower.
- Even with less manpower, law enforcement is expected to maintain the same level of services.
- Law enforcement must find innovative ways to fight crime through data systems.
- Big data can meet all of the above requirements if used to its fullest potential.

Law enforcement is the foundation upon which our society is kept safe, such that people can feel secure walking down the street in broad daylight, and a young woman can walk safely in remote areas at night and not feel threatened.

Lately, the public view of law enforcement has become negative. Police brutality is perceived to be on the rise. Crime seems to be escalating around the nation. Where is law enforcement when you need it the most? We must realize that law enforcement is in a catch 22 situation here in the 21st century. How can they be expected to do more with fewer resources? Are the demands being put on them improbable for them to realize?

They have fewer personnel to cover the same amount of ground as before, when law enforcement agencies had full rosters to meet all the demands of curtailing criminal activity.

Data Analytics is the Solution to Law Enforcement's Current Dilemma of Shrinking Resources

Once the previously mentioned data platform is put into place, then data analytics can be applied at a very detailed level. Police departments can now solve crimes faster than ever before. They can stop criminal activity with a greater success rate. They can even use data analytics to determine potential criminal behavioral tendencies and stop these lawbreakers from committing the crime before they even get the chance. Here are some technologies Microsoft uses to help in this quest:

- Microsoft Azure Data Lake
- Microsoft Cortana Analytics
- Windows speech recognition

There is one police department working with Microsoft and using their data analytic tools. Some of these tools are Microsoft Azure, which allows crime analysts to store criminal data in any size, shape, or speed. The beauty of this technology is that it can be used for any type of analytics or processing across platforms and languages. No matter how much data they have, it can all be stored in one central data system. Do you realize the significance of this?

Law enforcement agencies worldwide can network with each other. This would be helpful in tracking terrorist activities. By using predictive analytics, data can be extracted and used to project where terrorists will strike next. They can decipher the most likely form of attack the group or individuals will use. They can determine the extent of the operation based on historical data stored from previous attacks. They can stop the attacks before they materialize.

Microsoft Cortana Analytics: This analytics application allows crime analysts to predict crime scenarios using machine-learning algorithms at lightning-quick rates of speed. The analysts can also use this tool to support and enhance decision-making processes by recommending the best course of action to take in the event of a real-time crime or predictive crime taking place. It also helps them to automate and simplify critical decisions, which include many criminal variables that constantly change in real time, thus giving them a greater capability to make the right decisions more quickly to stop crimes before they happen and while they are happening.

Windows voice recognition: This assists police officers in dictating critical commands pertaining to a driver's outstanding warrant history, DUI violations, and ticket violation patterns. Knowing this information quickly in real time allows the police officer to position himself in a place to stop potential crimes or traffic violations from occurring. Essentially, the police officer could reduce the threat or actually diffuse a life-threatening event before it occurs.

Microsoft: A Central Player in Police Department Crime Prevention

Microsoft is using their data applications in assisting police departments to prevent crime before it happens. There is one police department they are working with to measure public safety in real time using varied sources. This particular police department uses a plethora of data streams to analyze live 911 calls, live feeds from cameras, and police reports to fight terrorism and other types of violent crimes. Crimes at all levels can be measured, which allows for quicker reaction time and provides the ability to develop strategies to fight these crimes in real time to proactively stop them before they happen.

Advanced Analytics used for Deciphering across-the-board Criminal Activities

Police departments are also using advanced analytics to automatically interpret crime patterns, such as burglaries, homicides, and domestic violence incidents. This data tool helps police departments quickly sort through large amounts of data. Police can now identify a burglar's crime patterns. When will the burglary will be committed, and what will be the point of entry into the house or building? What area will the burglar next try to steal from? What objects will he steal? Armed with this information, police officers can be positioned to stop the burglaries from happening.

Another facet is to prevent burglaries from happening in certain geographical areas and in certain timeframes. Let's say a certain area of the city is vulnerable to burglaries in certain times of the year. For example, a row of apartment complexes will be hit in the summer months between 8:00 a.m. and 12:00 p.m. With this knowledge, police departments can dispatch police officers to the designated areas during the crucial time period. Therefore, they can stop a series of robberies before they start.

Police Departments can Respond to Major Terrorist Crimes Successfully after they Occur

Remember the Boston Marathon bombing? Police used a few data sources to respond to the crime promptly and make arrests quickly. The Boston police used social media outlets to gather huge amounts of citizen data. It was a huge task to filter through the data, but it sure paid off. The police were able to identify and track down the two suspects.

Cellphone videos and security camera footage were painstakingly pored through, but the police's task paid off. The police were able to identify the two suspects in the video. They tracked them down from the images caught on these two data sources.

Chapter 25: Use of Big Data in the Public Sector

There are very few areas of life nowadays where using big data analysis will not be of benefit. We have already discussed how big data is being used by many industries, such as finance, travel, and law enforcement, so let us have a look at how other public sector agencies can benefit from the use of big data analysis.

Hospitals and ERs across the world always face one major issue, and this is unpredictability. This is why such institutions always need to have a large number of staff present to be able to cope with any unpredictable circumstances. However, even things like inflow of patients into an ER are not actually completely unpredictable. As time passes, even such random events start to show patterns, and this kind of data is also subject to data analysis with the use of some modern methods.

Hospitals around the USA have been using data analysis to predict daily patient inflow in recent years, and this kind of analysis has allowed them to predict the emergency admissions with 93% accuracy on any given day of the year. Both the patients and the staff of these hospitals benefit from this kind of data, as it becomes easier to allocate beds for non-emergency cases, and beds are less likely to need to be emptied for emergency cases. The data from such analysis also saves money, allows for fewer overtime hours, and creates a generally more proactive environment in which everyone can know what to expect rather than be on their toes all day long, waiting for things to happen and then reacting to them.

Other public offices have also been using big data to make their jobs easier and to achieve better results. For instance, the Australian Taxation Office has been using data analysis to look through massive amounts of taxation data and find possible tax fraud cases or small online retailers who are not meeting their obligations. There is a huge amount of ways tax data can be analyzed to discover all sorts of fraud, as well as trends among consumers.

School systems are another area of everyday life where big data is able to contribute greatly. Schools in all modern societies have started collecting big data and analyzing it to try and make various prognoses for the future. Data, such as student scores, levels of anxiety, or interest in particular subjects, can be used to discover all sorts of things. If such data is collected on a national level and analyzed properly, we could potentially be looking at numbers that will help the Department of Education improve the entire education system across the country in countless ways, thus saving budget dollars and improving education for the children. The privacy of such data has, however, been a major issue and one that is often debated as parents and children may see it as an intrusion into their privacy when such data is used for extensive analysis.

U.S. Government Applications

Big data has been playing a big role in the government of the USA for many years. The government owns six of the 10 most powerful supercomputers in the world, and these are mostly used for data analysis. It has been well-publicized that president Barack Obama used big data as a major part of his re-election campaign in 2012, after which his administration also announced the Big Data Research and Development Initiative, the goal of which it is to determine what kind of applications big data research may have in helping the government combat the current problems the nation is facing.

The NSA is currently also in the process of constructing possibly the world's biggest data center in Utah. It is supposed to be able to store several exabytes of data and will serve to analyze some of the most detailed and sensitive information in the world.

Big Data Security

As previously mentioned, the security of data collected by public agencies has become a major issue that has stirred up considerable debate. Ever since it has become known that agencies, such as the NSA, are collecting incredible amounts of information on almost everyone, the public has become enraged and questioned how exactly this data is being used and why it is being collected in the first place.

While big data indeed has the potential to be misused, the reality is that this fear often comes from ignorance more than reality. People outside of the industry tend to observe big data as a form of the "Big Brother" concept – a watchful eye that sees everything and has full control. However, many representatives of government agencies and experts in the field have warned the public that it is actually the private sector that is watching your every step, as private companies tend to be much more advanced in big data analysis and usually have much more information about the general public and particular individuals.

While some of the data collected by the public sector could possibly be misused, most of such data comes from "illegal" things, such as surveillance and other methods, some of which were spoken against by Edward Snowden, who leaked just how much information NSA had been collecting about the general public. The private sector, on the other hand, tends to collect vast amounts of information legally by having you voluntarily enter it into various agreements, sheets, and sign up forms online.

Public Sector Big Data Problems

Government agencies have major issues in using big data analysis. These agencies tend to obtain extremely large amounts of data to analyze, and because the average public sector employee is much older and less technologically savvy than an average private sector employee, it can be difficult to complete such contracts with precision and in a timely fashion.

Governments are unable to offer the competitive terms that private companies can, and for that reason, they are unable to get true experts to work for them. Thus, the end up employing less competent teams of data analysts. This human capital issue is one of the biggest problems that the public sector faces in analyzing big data. Private companies find various ways of legally sharing their big data and analysis, whereas the public sector tends to lag behind in this area.

Chapter 26: Big Data and Gaming

The gaming industry is a fairly wide term. When we say gaming, we encompass everything from social media games like Candy Crush Saga that are played by millions of Facebook users to Casino games like Roulette and Black Jack that are in core gambling games designed to provide entertainment though real money wins and losses.

Whether we are talking about innocent games that your 4-year-old could play or serious casino games where fortunes can be won or lost in a matter of hours, the gaming industry has taken to big data in a big way to improve their bottom lines. The gambling industry has been using numbers and data from the very start, as all casino games are basically number games in which the house will always win in the long term simply because the statistics say so. Gaming companies having nothing to do with gambling, such as Electronic Arts, have started using numbers as their main weapon in the age of information.

There are numerous ways for the gaming industry to improve their revenue by applying big data analysis. Everything from improving gaming experience, providing a more personalized gaming platform, and keeping the players playing longer can be achieved through extensive data analysis.

Improving the User Experience

There used to be a time when game manufacturers could only make a game, start selling it, and see how it does. They could then potentially go out and ask players what they thought of it and how they would like to improve it, but there was not much tangible data that could be used.

In recent years, much of all the gaming done around the world has went online. From social games played on the social networks to popular internet games like World of Warcraft, Counter Strike, and DOTA, players are now providing the game manufacturers with countless terabytes of useful information.

What kind of information are we giving away? Well, we can start with the timing. Companies are monitoring when average customers start playing a game, when they finish, how long they play, and other similar factors. Additionally, the game manufacturer will get data such as with whom you interacted in the game, where you played from, whether you used your smartphone or your laptop to play, how exactly you interacted within the game, and what kind of a goal you were out to complete. The list goes on and on, and there is no limit to the amount of information companies can gather about their players.

Eventually, like all data analyses, your gaming data turns into inferences. Based on a mix of all these factors, a game designer can improve your personal gaming experience, whether it is by adding more content that actual players enjoy or by awarding you in various ways within the game. If a game is becoming popular on smartphones, they could make a patch to make it more mobile friendly, and if people enjoy playing it on laptops, they may decide to make it look better on bigger screens. There is simply very little that gets away from game manufacturers in the modern day, with all the data they are able to collect on their end users.

Various games have used big data to find major issues within their games. Thus, their results have improved dramatically. For instance, a company could, through the analysis of big data, find out that a certain early level of their game is giving people too much trouble, causing them to quit. In turn, they can make this level easier, thus allowing players to remain in the game longer and increasing their revenue by doing so.

Social games have been selling things like player items, additional content, or chips to play with. Using big data can tell the companies exactly when, how, and why people choose to purchase or not purchase these items, and they can then use this information to improve the sales of in-game items.

Big Data in Gambling

As previously mentioned, the gambling industry is on the forefront when it comes to using big data to make their customers spend more money on their products. Many people see the gambling industry as some kind of a devil, but it is, in fact, just another part of the entertainment industry, and like others, it does tend to use various tactics to make users leave their money. After all, there is little difference between making a customer lose money by playing Black Jack and making them buy a product they will never use in their lives.

The first to adopt big data in the gambling industry was the bookmakers. As their job is to make the odds for sporting events appealing enough to their users that they would place a bet and yet low enough that they can turn a profit, real-time big data analysis was exactly what they have been yearning for. When the software became available to do this, live betting on sporting events boomed, and the companies could be sure to stay a step ahead of their customers as they simply had more data available to them.

By using big data, bookmakers have been able to predict the outcomes of various events with more certainty than any man could do on his own. Processing data from hundreds or thousands of matches played by a sports team in the past allows bookies to have an edge over an average customer. Processing the odds of live matches also allows them to adjust the odds of events automatically, thus avoiding the off chance of a diligent punter seeing something they have missed.

It did not take much time for the other side to retaliate, however. If betting companies could predict the outcomes of matches, many bettors thought to themselves, why can't we do the same, but better? Major internet giants, such as Google and Yahoo, actually used big data analytics to predict the outcomes of matches and, in doing so, showed that there may be a method to the madness. In a famous case, Google got 14 out of 16 and Yahoo 15 out of 16 match outcomes at the 2014 Football World Cup correctly. While these may have mostly been the favorites winning, the numbers were still pretty impressive, albeit limited in many ways. Many punters have also been trying to use big data to give themselves an edge, and the reality is that we may never find out how successful they are as this is not the kind of information anyone would want to share.

However, predicting the odds of events is not the only way gambling companies have been using big data to improve their bottom line. Brick-and-mortar casinos, such as Harrah's and Caesar's Palace, have long been using big data analysis to find out information about their customers, and they use this information to get these customers to gamble more.

For instance, casinos track the amount of time players spend at a slot machine, the moment they get annoyed enough with the game to leave, and what kind of slot games makes them stay and play the longest. By examining this type of data, casinos can improve their marketing strategies, place different machines and games better around the floor, or provide players with promotions and incentives to counter various negative outcomes, such as a player leaving the casino or not returning. In a famous example, Caesar's Palace managed to increase the number of returning customers dramatically by awarding losing players with free dinner coupons on their way out of the casino. By doing so, they were able to increase their revenue significantly, and it was all thanks to big data.

In a similar manner as supermarkets have used big data analysis to better place their products, large Las Vegas casinos have adopted this technique. In fact, when you are walking the floor of any major casino, you are very likely looking at a well-designed and carefully planned maze that was designed based on how customers move around the casino floor. The placement of every single slot machine has actually been designed to optimize revenue.

Beating the System

As previously discussed, the moment big data analysis started being used by betting companies, the other side representing the players started to respond. Today, there are many websites that use big data to help gamblers gamble smarter, and while some of these may easily be a complete sham, others have been pretty reliable and useful tools for gamblers to reduce their losses or even to win.

Poker is a fantastic example of a game where numbers completely rule, and players with a good understanding of mathematics and statistics have an edge over other players. Given that the game is played exclusively between players and the house has no profit from it, this table game has actually been one where a significant number of players have been able to win on a consistent basis.

However, such software as Sharkscope and Hold'em Manager, which were created for poker players, actually took this one step further. Sharkscope is an online service that allows players to examine the results of hundreds of thousands of online tournament players in real time, providing them with such statistics as average buy in, return of investment, and other useful data. To a trained eye, this type of information can mean the difference in spotting the target at the tables, and the tool has been vastly popular among poker players.

Analyzing data at the table itself is where it gets truly interesting. Tools like Hold'em Manager and Poker Tracker have enabled poker players to keep track of their opponents' statistics in terms of the way they play their hands. By storing data from every hand played at the table and every hand played against a particular opponent, these tools have been able to make fantastic inferences that have allowed professional players to absolutely dominate at the tables for many years.

Naturally, sport betting was also targeted by those looking to make a profit, and it is well-known that sports bookies get their info based on big data. As such, you do not really need to know the results; you just need to know how to predict them better than the bookies. This is where sites, such as Betegy, have stepped in, trying to beat the odds and promising to provide their customers with relevant data that could help them beat the bookmakers. This particular website claims they can guess the outcomes of 90% of English Premier League games and, if true, it is truly revolutionary. Whether the sites can truly predict results at the moment remains a mystery, but the reality is that big data analysis may mean an end to many of the current types of gambling as an increasing number of software like this one are developed and as they become more accurate.

The Growth of Gaming

While the gambling industry has been huge for decades, the video game industry has only recently seen its rise to stardom. Thousands of people have been paying entry fees to various gaming events and tournaments for such games as League of Legends. These games have been attracting a huge viewership and million-dollar prize pools. In fact, the gaming industry is valued today at over $100 billion, and it is still rapidly growing.

This is the exact reason why gaming companies, such as Electronic Arts, have been spending huge amounts of money on acquiring and analyzing the big data collected from their players. The competition has never been fiercer, and gamers only have a limited amount of time and money they can spend on video games.

With so much competition out there, gamers today are looking for excellence, both in game design and in customer personalization and service, which is why big data has never been more useful. With such a rapidly expanding market, grabbing new and keeping old customers are paramount.

What better way is there to make a good game that will sell well than listening to the gamers themselves? Given that there is now so much data available for the companies to look into, all it takes is proper analysis of such data. If they are able to analyze the data in the right way, it is nearly impossible to make a game that will fail. Conversely, poor analysis can lead to fiascos, so the amount of responsibility resting on data scientists in the gaming industry is immense.

While you may have thought that the internet connectivity of your Xbox or PlayStation was just for your sake, think again. Companies are actually collecting data on the way you use the consoles, even when you are not playing online. Every hour of gaming time you log is recorded if your device is connected to the internet, and this type of information is used when creating new games.

Social media games present an even bigger challenge than normal games as they do not sell but are free to play. Gaming companies use the big data collected to find ways of monetizing these games, and this is a complex process that involves many factors.

In the end, it is sufficient to say that the gaming industry is one of those that have benefited the most from the introduction of big data analysis into the business world, and you can expect the games to move more and more toward what the users want to see as more big data is crunched every single day.

Chapter 27: What about Prescriptive Analytics?

What is Prescriptive Analytics?

The latest technical data buzz word, "Prescriptive Analytics," is the third branch of the big three data branches. It is predicted that this analytic model will be the last of the three. Is it the next groundbreaking data model in technology? This analytics system is a summation of the previous two, namely, descriptive analytics and predictive analytics. Below are the definitions of the three analytic models:

- **Descriptive Analytics** – uses data aggregation and data mining techniques to give enlightenment to the past. It answers the question, "What has happened?"
- **Predictive Analytics** – uses statistical models and forecasting techniques to giving insight into the future. It answers the question, "What will happen?
- **Prescriptive Analytics** – uses optimization and simulation algorithms to determine options for the future. It answers the question, "What should we do?"

The prescriptive analytics model is the newest branch in the data world. Some are already arguing that it is the only way to go in the analytical world in the future. It doesn't seem like prescriptive analytics hasn't caught fire in the business world at present. It will become more widespread in different industries. Companies will realize the benefits of having a model that will suggest answers to future questions, which is the most advanced component of this relatively new technology.

This takes predictive analytics one step further. It goes beyond predicting what will happen in the future. It gives options for the future and then suggests the best option for that course of action. It has some "artificial intelligence" involved in its processes. It analyzes optimization and simulation algorithms and then "thinks" of all the options based on the systems of these algorithms. It will even suggest the best option to take out of all the options analyzed. Prescriptive analytics is, comparatively speaking, more difficult to administer than the other two analytical models. It also uses machine thinking and computational modeling.

These techniques are used against several data sets, including historical data, transactional data, real-time data feeds, and big data. Prescriptive analytics also incorporates algorithms that are released on data with minimal parameters to tell organizations what to do. The algorithms are programmed to conform to the changes in established parameters instead of being controlled externally by humans. The algorithms are allowed to be optimized automatically. Over time, their ability to predict future events improves.

What are the Benefits of Prescriptive Analytics?

Many companies aren't currently using prescriptive analytics in their daily operations. Some huge corporations are using it in their daily operations. These big companies are effectively using prescriptive analytics to optimize production and successfully schedule and guide inventory in their respective supply chains to be sure that they are delivering the right products at the right times and to the right places. This also optimizes the customer experience. It is advised to use prescriptive analytics to tell customers what to do. This can be a very effective management tool for end-users.

The Future of Prescriptive Analytics

What does the future hold for prescriptive analytics in the international business world? In 2014, approximately 3% of the business world was using prescriptive analytics. Scott Zoldi, Chief Analytics Officer at FICO, made the following predictions concerning prescriptive analytics for 2016:

- The cornerstone of prescriptive analytics, "streaming analytics," will become dominant in 2016.
- Streaming analytics is to be applied to transaction-level logic for real-time events (They have to occur in a prescribed window, such as last 5 seconds, last 10,000 observations, etc.).
- Prescriptive analytics will be a must-have cyber security application (This process will analyze suspicious behavior in real time).
- Prescriptive analytics will become mainstream in lifestyle activities in 2016 (From home appliances to automated shopping, it is poised for explosive growth).

Analytical experts are predicting the next major breakthrough for prescriptive analytics to be that it will go mainstream, spreading across all industries. In 2015, it was predicted that prescriptive analytics would increase in relation to consultancy partnering with academics. As of 2016, there still was no off-the-shelf general purchase for prescriptive analytics, and this seems to be a long way off.

It seems to be a hit-and-miss proposition for prescriptive analytics within departments of the same company. One department will be using prescriptive analytics, while other departments wouldn't be.

Google using Prescriptive Analytics for their "Self-Driving Car"

The search engine giant is using prescriptive analytics in their five-year-old self-driving cars. The cars drive automatically without a human guide. They make lane changes, turn right or left, slow down for pedestrians, and stop at lights. Generally, they drive like any other car, but without the human driver.

The car uses prescriptive analytics for every decision it will need to make during the driving experience. It makes decisions based on future events. For instance, when the car approaches a traffic light intersection, it has to decide on whether to turn right or left. It will take immediate action on future possibilities concerning which way to turn. It considers many factors before it makes that decision. It considers what is coming towards it in terms of traffic, pedestrians, etc., before it determines which way to turn. It even considers the effect of a certain decision before determining whether to turn right or left at the intersection. Google, being such a high profile multi-billion-dollar company, will have a major impact on prescriptive analytics moving forward across the business world.

The Oil and Gas Industry is using Prescriptive Analytics

The oil and gas industry is currently using prescriptive analysis to interpret structured and unstructured data. It also uses this analytical model to maximize fracking (i.e., the process of injecting liquids at high pressure to subterranean rock, boreholes, etc., to force open existing fissures to extract oil and gas). Below are applications of using prescriptive analytics in the oil and gas industry:

- Maximize scheduling and production, as well as tune the supply chain process.
- Maximize customer experience.
- Locate functioning and non-functioning oil wells.
- Optimize the equipment materials needed to pump oil out of the ground.
- Accomplish all the above tasks to deliver the right products to the right customers at the perfect time.

Most of the world's data (about 80%) comprises unstructured information, such as videos, texts, images, and sounds. The oil and gas industry used to look at images and numbers, but in separate silos. The advent of prescriptive analytics changed this. Now, the industry has the ability to analyze hybrid data, that is, the combination of structured and unstructured data, which gives the industry a much clearer picture and more complete scenario of future opportunities and problems. This, in turn, gives them the best actions to determine more favorable outcomes. For example, to improve the hydraulic fracking process, the following datasets must be analyzed simultaneously:

- Images from well logs, mud logs, and seismic reports
- Sounds of fracking from fiber optic sensors
- Texts from drillers and frack pumpers' notes
- Numbers from production and artificial lift data

Along these lines, hybrid data is essential for analysis because of the multi-billion-dollar investment and drilling decisions made by oil companies. They must know where to drill, where to frack, and of course, how to frack. If any one of these steps is skipped or done incorrectly, it could have disastrous results for these energy companies. The cost could be astronomical. It would be unimaginable and a waste of time, energy, and manpower if they fracked or drilled in the wrong way or in the wrong place.

There is more to the components involved in completing prescriptive analytics. Scientific and computational disciplines must be combined to interpret different types of data. For instance, to algorithmically interpret images like log wells, machine thinking must be connected with pattern recognition, computer vision, and image processing. Mixing these disciplines enables energy companies to visualize a more holistic system of recommendations, such as where to drill and frack, while minimizing issues that might come along the way. By forming detailed traces and using data from production, subsurface, completion, and other sources, energy companies are able to determine the functioning and non-functioning wells in any field.

This process is anchored by prescriptive analytics

technology's ability to digitize and understand well logs to produce dispositional maps of the subsurface. Once they know where to drill, oil companies save untold millions in resources. This process helps them to pinpoint where non-functional wells are located, obviously saving them optimum drilling time. Notably, the environmental impact is minimized on the particular landscape. Prescriptive analytics should be used in other areas of oil and gas production. In both traditional and non-traditional wells, simply by using data from pumps, production, completion, and subsurface characteristics, oil companies are able to predict the failure of submersible pumps and thus reduce production loss, saving untold millions of dollars.

The Apache Corporation is using prescriptive analytics to predict potential failures in pumps that extract oil from the subsurface, thereby minimizing the associated costly production loss attributed to the failure of these pumps. Another potential application of prescriptive analytics is to hypothetically predict corrosion development on existing cracked oil pipelines. The prescriptive model could describe pre-emptive and preventive actions by analyzing video data from cameras. It could also analyze additional data from robotic devices known as "smart pigs," which are positioned within these damaged pipelines.

Using prescriptive analytics in their daily operations will help oil companies make wiser decisions. This will result in fewer manpower accidents, lower production loss, maximized resources, and increase financial profits. Another valuable benefit is the reduction of the environmental impact on particular landscapes and surrounding communities. Oil companies will place themselves in a favorable light with environmentalists by showing concern and taking steps to minimize or reduce environmental hazards around their operating facilities.

Hopefully, the oil companies will deem it fit to pass on the cost-savings to the consumers in the form of reduced gas prices by using the prescriptive analytics model. Further investigation into the oil and gas industry reveals that they use prescriptive analytics in deeper ways. They use it to locate the oil fields with the richest concentrations of oil and gas. It helps track down oil pipes with leaks. When caught early, these leaks can be repaired, thus avoiding a major environmental hazard from an oil spill. Prescriptive analysis also helps refine the fracking process to avoid endangering the environment and to improve the output of oil and gas.

Conversely, the oil and gas industry will have to increase expenses to hire the proper experts to run the prescriptive analytics models. The training involved in getting these technicians up to speed with company operations will further inflate company costs. Nevertheless, streamlining all the processes outlined earlier in this chapter will outweigh the increased costs required to increase manpower to further justify that the hiring of analysts, statisticians, and data scientists is a necessary expense. The prescriptive analytics model is more complicated to implement than the descriptive and predictive analytical models. However, the potential to change industries is a given, as the prescriptive analytics system has already drastically changed the oil and gas industry.

Prescriptive Analytics is making Significant Inroads into the Travel Industry

The travel industry has come on board and is using prescriptive analytics in their daily activities, as well as in their overall operational functions. Prescriptive analytics calls for many large data sets. Given this factor, the travel industry sees great potential in this latest round of data analytics. Online traveling websites, such as airline ticketing sites, car rental sites, and hotel websites, have seen the tremendous benefits of predictive analytics in their local business functions.

They are implementing prescriptive analytics to filter through multiple and complex phases of travel factors, purchases, and customer factors, such as demographics and sociographics, demand levels, and other data sources to maximize pricing and sales. Predictive analytics may turn the travel industry upside down as it did the oil and gas industry. It could be said that predictive analytics is re-inventing the wheel of data analytics. It is more complicated in its daily interface than its two earlier predecessors.

The other applications of prescriptive analytics in the travel industry include segmenting through potential customers predicated on multiple data sets on how to spend marketing dollars. By doing this, the travel industry can optimize every dollar to attract a wider customer base. They will be able to predict what the customers' preferred traveling locations are based on past traveling preferences. This will result in marketing and advertising strategies that will appeal to the particular target age groups.

As an example, the Intercontinental Hotel Group currently utilizes 650 variables on how to spend their marketing dollars to increase their customer base.

Other Industries using Prescriptive Analytics

The healthcare industry is waking up to the cutting-edge technology that prescriptive analytics can provide to their intricate overall operational activities. Using so many variables offers doctors, nurses, and physician assistants the optimal alternatives for the best treatment programs for their patients. Also, the prescriptive analytics model can suggest which treatment would be the best fit for the illness or physical condition of the patient. This will streamline the diagnostic process for medical professionals. Imagine the hours saved because the many options for any medical condition will be populated through the numerous data sets used in the prescriptive analytics model.

One very successful example of a medical company saving a large amount of money is the case of the Aurora Healthcare Centre, which was able to improve healthcare and reduce re-admission rates by a healthy 10%, thus resulting in a significant savings of $6 million dollars. This is the type of potential prescriptive analytics has in impacting industries worldwide.

The pharmaceutical industry will not be left out of the loop either. Prescriptive analytics can help the industry across the board by reducing drug development and minimizing the time it takes to get the medicines to the market. This would definitely reduce drug research expenditures and greatly decrease manpower hours and resources. Moreover, drug simulations will shorten the time it takes to improve the drugs, and patients will be easier to find for trials on the new medications.

We can see that prescriptive analytics is the wave of the future in the data analytics world. However, the model does require massive amounts of data sets. Currently, only the largest corporations have the massive data sets needed to run the prescriptive analytics system feasibly. If the data sets are not available in great quantities, the system absolutely will not work. This is a drawback for smaller companies that do not have large quantities of data sets to effectively run prescriptive analytics in their operations. This could be one reason why this model has not taken off at this point. Nevertheless, the model is starting to make crossroads into the larger corporate setting as organizations realize the great potential this analytical model has. It could take five to 10 years for prescriptive analytics to become a household system in businesses at the international level.

The measurable impact that prescriptive analytics has had in the oil and gas and travel industries cannot be ignored. Truly, prescriptive analytics could be the dominant data system of the near future. It has made noteworthy crossroads thus far.

Data Analysis and Big Data Glossary

The following words and terms are related to data analysis, big data, and data science. While not all of them have been used in this book, these are words and terms that you will come across in your studies. The whole list of terms is quite exhaustive and would take forever to list, so I have chosen to give you definitions for the more common terms.

A

ACID Test –a test applied to data to test for consistency, durability, isolation, and atomicity. In other words, it makes sure it all works as it should.

Aggregation – the collection of data from a number of different databases for analysis or data processing.

Ad hoc Reporting – reports that are generated for one-off use.

Ad Targeting – an attempt made by a business to use a specific message to reach a specific audience. This usually takes the form of a relevant ad on the internet or by direct contact.

Algorithm – a mathematical formula that is inserted into software to analyze a certain set of data.

Analytics – the art of using algorithms and certain statistics to determine the meaning of data.

Analytics Platform – either software or a combination of software and hardware that provides you with the computer power and tools that you need to perform different queries.

Anomaly Detection – the process by which unexpected or rare events are identified in a dataset. These events will not conform to any other events in the same dataset.

Anonymization – the process of severing the links between people and their records in a database. This is done to prevent the discovery of the source of the records.

Application Program Interface (API) –programming standards and sets of instructions that enable you to access or build web-based software applications.

Application – a piece of software designed to perform a certain job or a whole suite of jobs.

Artificial Intelligence – the apparent ability of a computer to apply previously gained experience to a situation in the way that a human being would.

Automatic Identification and Capture (AIDC) – refers to any method by which data is identified and collected and then stored in a computer system. An example of this would be a scanner that collects data through an RFID chip about an item that is being shipped.

B

Behavioral Analytics – the process of using data collected about people's behavior to understand the intent and then predict future actions.

Big Data – there are many definitions for the term, but all of them are pretty similar. The first definition came from Doug Laney in 2001, then working for META Group as an analyst. The definition came in the form of a report called "3-D Data Management: Controlling Data Volume, Velocity, and Variety." Volume refers to the size of the datasets, and a McKinsey report, called "Big Data: The Next Frontier for Innovation, Competition and Productivity," goes further on this subject, stating that "Big data refers to datasets whose size is beyond the ability of typical database software tools to capture, store, manage, and analyze."

Velocity refers to the speed the data is acquired and used. Companies are not only collecting the data at a faster speed, but they also want to determine its meaning at a faster rate, possibly in real time.

Variety refers to the different data types available for collection and analysis, as well as the structured data that would be found in a normal database. There are four categories of information that make up big data:

- Machine generated – including RFID data, data that comes from monitoring devices, and geolocation data from a mobile device
- Computer log – including clickstreams from some websites
- Textual social media – from Twitter, Facebook, LinkedIn, etc.
- Multimedia social media –any other information gleaned from places like YouTube, Flickr, and other sites that are similar

Biometrics – the process of using technology to use physical traits to identify a person, i.e., fingerprints, iris scans, etc.

Brand Monitoring – the process of monitoring the reputation of your brand online, normally with the use of software.

Brontobyte – a unit representing a vast amount of bytes. Although it is not yet officially recognized as a unit, a brontobyte has been proposed as a unit measure for data that goes further than the yottabyte sale

Business Intelligence (BI) – term that is used to describe the identification of data. It encompasses extraction and analysis as well.

C

Call Detail Record Analysis (CDR) –contains data collected by telecommunications companies and includes the time and the length of a phone call. The data is used in a variety of different analytical operations.

Cassandra – one of the most popular columnar databases normally used in big data applications. Cassandra is an open-source database that is managed by The Apache Software Foundation.

Cell Phone Data – a mobile device is capable of generating a vast amount of data, most of which can be used in analytical applications.

Classification Analysis – data analysis used for assigning data to specific groups or classes.

Clickstream Analysis – the process of analyzing web activity by collecting information about what a person clicks on a page.

Clojure –a dynamic program language that is based on LISP and uses JVM (Java Virtual Machine); suitable for use in parallel data processing.

Cloud – a wide term covering any service or web-based application hosted remotely.

Clustering Analysis – the process of analyzing data and using it to identify any differences or similarities in data sets so that any that are similar can be clustered together.

Columnar Database/Column-Oriented Database – a database that stores data in columns, not rows. Row databases could contain information like name, address, phone number, age, etc., all stored on one row for one person. In column databases, all of the names would be stored in one column, addresses in another, ages and telephone numbers in their own columns, and so on. The biggest advantage of this type of database is that disk access is faster.

Comparative Analysis – analysis that compares at least two sets of data or processes to see if there are any patterns in large sets of data.

Competitive Monitoring – the use of software to automate the process of monitoring the web activity of competitors.

Complex Event Monitoring (CEP) –the process of monitoring the events in the systems of any organization, analyzing them, and then acting when necessary.

Complex Structured Data – structured data made up of at least two inter-related parts, thus making it difficult for structured tools and query languages to process.

Comprehensive Large Array-Data Stewardship System (CLASS) – digital library of historical data gained from NOAA (US National Oceanic and Atmospheric Association).

Computer-Generated Data – data that is generated through a computer instead of a human being; for example, a log file.

Concurrency – the ability to execute several different processes at once.

Confabulation – the process of making a decision based on intuition as if it were based on data instead.

Content Management System (CMS) – software that allows for the publication and the management of web-based content.

Correlation Analysis – the process of determining the statistical relationship between two or more variables, usually to try to identify if there are any predictive factors present.

Correlation – refers to multiple classes of dependent statistical relationships. Examples would be the correlation of parents and their children or the demand for a specific product and its price.

Cross-Channel Analytics – analysis that can show lifetime value, attribute sales, or an average order.

Crowdsourcing – the process of asking the public to help complete a project or find a solution to a problem.

Customer Relationship Management (CRM) – software used to help manage customer service and sales.

D

Dashboard – graphical report of either static or real-time data on a mobile device or a desktop. The data is usually high level in order to give managers access to quick reports of performance.

Data – a qualitative or quantitative value. Examples of more common data include results from market research, sales figures, readings taken from monitoring equipment, projections for market growth, user actions on websites customer lists, and demographic information.

Data Access – the process of retrieving or viewing data that has been stored

Digital Accountability and Transparency Act 2014 (DATA Act) – a relatively new U.S. law that is intended to make it easier to gain access to federal government expenditure information by requiring the White House Office of Management and Budget and the Treasury to standardize data on federal spending and to publish it.

Data Aggregation – the collection of data from a number of sources for the purpose of analysis or reporting.

Data Analytics – the use of software to determine the meaning of information from data. The result may be a status indication, a report, or an automatic action based on the information received.

Data Analyst – the person who is responsible for preparing, modeling, and cleaning data so that actionable information can be gained from it.

Data Architecture and Design –the structure of enterprise data. The actual design or structure will vary, as it is dependent on the result that is required. There are three stages of data architecture:

- The conceptual representation of the entities in the business
- The logical representation of the relationship between each of the entities
- The construction of the system that supports the functionality

Data Center – physical place that is home to data storage devices and servers that may belong to one or more organizations.

Data Cleansing – the review and revision of data to eliminate any duplicate information, to correct spelling errors, to add any missing data, and to provide consistency across the board.

Data Collection – a process that captures any data type.

Data Custodian – the person responsible for the structure of the database and the technical environment, including data storage.

Data Democratization – the concept of ensuring that data is directly available to all workers in an organization, instead of them having to wait for the data to be delivered to them by another party, usually the IT department, within the business.

Data-Directed Decision Making – the use of data to support the need to make crucial decisions.

Data Exhaust – the data created by a person as a

byproduct of another activity; for example, call logs or web search histories.

Data Feed – the means by which data streams are received; for example, Twitter or an RSS feed.

Data Governance – the rules or processes that ensure data integrity and guarantee that management best practices are followed and met.

Data Integration – the process by which data from various sources is combined and presented in one view.

Data Integrity – a measure of trust that an organization places in the completeness, accuracy, validity, and timeliness of data.

Data Management – the Data Management Association states that data management should include these practices to ensure that the full lifecycle of data is managed:

- Data governance
- Data design, analysis, and architecture
- Database management
- Data quality management
- Data security management
- Master data management and reference
- Business intelligence management
- Data warehousing
- Content, document and record management

- Metadata management
- Contact data management

Data Management Association (DAMA) – a non-profit international organization for business and technical professionals that is "dedicated to advancing the concepts and practices of information and data management."

Data Marketplace – an online location where people can purchase and sell data.

Data Mart – access layer of a data warehouse that provides users with data.

Data Migration – the process by which data is moved between different formats, computer systems, or storage places.

Data Mining – the process of determining knowledge or patterns from large sets of data.

Data Model/Modeling – data models are used to define the data structure needed for communication between technical and functional people to show which data is needed for the business. It is also used for the communication of development plans for data storage and access among specific team members, usually those in application development.

Data Point – individual item on a chart or graph.

Data Profiling – the collection of information and statistics about data.

Data Quality – the measurement of data to determine if it can be used in planning, decision making, and operations.

Data Replication – the process by which information is shared to ensure consistency between sources that are redundant.

Data Repository – the location where persistently stored data is kept.

Data Science –a relatively recent term that has several definitions. It is accepted as a discipline that incorporates computer programming, data visualization statistics, data mining database engineering, and machine learning in order to solve complex problems.

Data Scientist – a person who is qualified to practice data science.

Data Security – the practice of ensuring that data is safe from unauthorized access or from destruction.

Data Set – a collection of data stored in tabular form.

Data Source – a source from which data is provided, such as a data stream or database.

Data Steward – a person responsible for the data that is stored in a data field.

Data Structure – a certain method for the storage and organization of data.

Data Visualization – visual abstraction of data that is used for determining the meaning of the data or for communicating the information in a more effective manner.

Data Virtualization – process by which different sources of data are abstracted through one access layer.

Data Warehouse – a place where data is stored for

analysis and reporting purposes.

Database – digital collection of data and the structure that the data is organized around.

Database Administrator (DBA) – a person who is usually certified for being responsible for the support and maintenance of the integrity of content and structure of a database.

Database as a Service (DaaS) – a database that is hosted in the cloud and is sold on a metered basis. Examples of this include Amazon Relational Database Service and Heroku Postgres.

Database Management System (DBMA) – software used for collecting and providing structured access to data.

De-Identification – the process of removing data that links specific information to a specific person.

Demographic Data – data that relates to the characteristics of the human population, i.e., in a specific area, age, sex, etc.

Deep Thunder – weather prediction service from IBM that provides specific organizations, such as utility companies, with weather data, thus allowing them to use the information to optimize the distribution of energy.

Distributed Cache – data cache that spreads over a number of systems but works as a single system, typically used as a way of improving performance.

Distributed File System – a file system that is mounted on several servers at the same time to enable data and file sharing.

Distributed Object – software module that has been designed to work with other distributed objects that are stored on different computers.

Distributed Processing – execution of a specific process over several computers that are connected to the same computer network.

Document Management – the process of tracking electronic documents and paper images that have been scanned and then storing them.

Drill – this is an open-source system distributed for carrying out interactive analysis on very large datasets.

E

Elastic search – open-source search engine that is built on Apache Lucene.

Electronic Health Records (EHR) – a digital health record that should be accessible and usable across a number of different healthcare settings.

Enterprise Resource Planning (ERP) – software system that enables a business to manage resources, business functions, and information.

Event Analytics – shows the steps taken to lead to a specific action.

Exabyte – 1 billion gigabytes or one million terabytes of information.

Exploratory Data Analysis – data analysis approach that focuses on identifying general data patterns.

External Data – data that lives outside a system.

Extract, Transform, and Load (ETL) – process used in data warehousing to prepare data for analysis or reporting.

F

Failover – the process of automatically switching to a different node or computer when one fails.

Federal Information Security Management Act (FISMA) – U.S. federal law that states that all federal agencies have to meet specific information security standard across all systems.

Fog Computing – computing architecture that enables users to have better access to data and data services by moving cloud services like analytics, storage, and communication closer to them through a device network that is distributed geographically.

G

Gamification – the process of using gaming techniques in applications that are not games. This is used to motivate employees and encourage specific behaviors from customers. Data analytics often is applied to this in order to personalize rewards and encourage specific behaviors to get the best result.

Graph Database – a NoSQL database that makes use of graph structures for semantic queries with edges, nodes, and properties that store, query, and map data relationships.

Grid Computing – performance of computing functions that make use of resources from systems that are multi-distributed. The process involves large files and is mostly employed for multiple applications. The systems that make up a grid network do not need to be designed similarly; neither do they have to be in the same location geographically.

H

Hadoop – open-source software library that is administered by Apache Software Foundation. Hadoop is defined as "a framework that allows for the distributed processing of large data sets across clusters of computers using a simple programming model."

Hadoop Distributed File System (HDF) – a fault-tolerant distributed file system that is designed to work on low-cost commodity hardware. It is written for the Hadoop framework and is in the Java language.

HANA – a hardware/software in-memory platform that comes from SAP. It is designed for real-time analytic and high volume transactions.

HBase – distributed NoSQL database in columnar format.

High-Performance Computing (HPC) – also known as supercomputers. Normally, these are custom-built using state of the art technology as a way of maximizing computing performance, throughput, storage capacity, and data transfer speeds.

Hive – a data and query warehouse engine similar to SQL.

I

Impala – open-source SQL query engine distributed for Hadoop.

In-Database Analytics – the process of integrating data analytics into the data warehouse.

Information Management – the collection, management, and distribution of all different types of information, including paper, digital, structured and unstructured data.

In-Memory Database – a database system that relies solely on memory for storing data.

In-Memory Data Grid (IMDG) – data storage in memory across a number of servers for faster access, analytics, and bigger scalability.

Internet of Things (IoT) – the network of physical things that are full of software, electronics, connectivity, and sensors to enable better value and service through the exchange of information with the operator, manufacturer, and/or another connected device/s. Each of these things is identified through its own unique computing system, but is also able to interoperate within the internet infrastructure that already exists.

K

Kafka – open-source message system from LinkedIn that is used to monitor events on the web.

L

Latency – a delay in the response from or delivery of data to or from one point to another.

Legacy System – an application, computer system of a technology that is now obsolete but is still used because it adequately serves a purpose.

Linked Data – this is described by Time Berners Lee, the inventor of the World Wide Web, as "cherry-picking common attributes or languages to identify connections or relationships between disparate sources of data."

Load Balancing – the process of the distribution of a workload across a network or cluster as a way of optimizing performance.

Location Analytics – this enables enterprise business systems and data warehouses to use mapping and analytics that is map-driven. It enables the use of geospatial information and a way to associate it with datasets.

Location Data – data used to describe a particular geographic location.

Log File – files created automatically by applications, networks, or computers to record what happens during an operation. An example of this would be the time that a specific file is accessed.

Long Data – this term was created by Samuel Arbesman, a network scientist and mathematician, referring to "datasets that have a massive historical sweep."

M

Machine-Generated Data – data that is created automatically from a process, application, or any other source that is not human.

Machine Learning – a process that uses algorithms to enable data analysis by a computer. Its purpose is to learn what needs to be done when a specific event or pattern occurs.

Map/Reduce – general term that refers to the process of splitting a problem down into bits, each of which is then distributed among several computers that are on the same cluster, network, or a map, that is, a grid of geographically separated or disparate systems. The results are collected from each computer and combined into a single report.

Mashup – a process by which different datasets from within one application are combined to enhance the output. An example of this would be a combination of demographic data with a real estate listing.

Massively Parallel Processing (MPP) – the process of breaking up a program into bits and executing each part separately on its own memory, operating system, and processor.

Master Data Management (MDM) –any data that is non-transactional and is critical to business operations, i.e., supplier or customer data, employee data, and product information. MDM is the process of master data management to ensure availability, quality, and consistency.

Metadata – data that is used to escribe other data, i.e., date of creation and size of a data file.

MongoDB – NoSQL database that is an open source under the management of 10gen.

MPP Database – database that is optimized to work in an MPP processing environment.

Multidimensional Database – a database that is used to store data in cubes or multidimensional arrays instead of the usual columns and rows used in relational databases. This allows the data to be analyzed from a number of angles for analytical processing and for complex queries on OLAP applications.

Multi-Threading – the process of breaking up an operation in a single computer system into several threads so it can be executed faster.

N

Natural Language Processing – the ability of a computer system or program to understand human language. Its applications include enabling humans to interact with a computer by way of automated translation and speech and by determining the meaning of unstructured data, such as speech or text data.

NoSQL – database management system that avoids using the relational model. NoSQL can handle large volumes of data that do not follow a fixed plan. It is best suited for use with large volumes of data that do not need the relational model.

O

Online Analytical Processing (OLAP) – the process of using three operations to analyze multidimensional data:

- Consolidation – aggregation of available factors
- Drill-down – allows users to see details that are underlying the main data
- Slice and Dice – allows users to choose subsets and see them from different perspectives

Online Transactional Processing (OLTP) – process by which users are given access to vast amounts of transactional data in a way that they are able to determine a meaning from it.

Open Data Center Alliance (ODCA) – a consortium of IT organizations from around the globe that has a single goal, i.e., to hasten the speed at which cloud computing is migrated.

Operational Data Store (ODS) – location used to store data from various sources so that a higher number of operations can be performed on the data before it is sent for reporting to the data warehouse.

P

Parallel Data Analysis – the process of breaking up an analytical problem into several smaller parts. Algorithms are run on each part at the same time. Parallel data analysis can happen in a single system or in multiple systems.

Parallel Method Invocation (PMI) – a process that

allows programming code to call many different functions in parallel.

Parallel Processing — the ability to execute several tasks at once.

Parallel Query – the execution of a query over several system threads to speed up performance.

Pattern Recognition – labeling or clarification of a pattern that is identified in the machine learning process.

Performance Management – process of monitoring the performance of a business or system against goals that are predefined to identify any specific areas that need to be monitored.

Petabyte – 1024 terabytes or one million gigabytes.

Pig – data flow execution and language framework used for parallel computation.

Predictive Analytics – the use of statistical functions on at least one dataset to predict future events or trends.

Predictive Modeling – the process by which a model is developed to predict an outcome or trend.

Prescriptive Analytics – the process by which a model is created to "think" of all the possible options for the future. It will suggest the best option to take.

Q

Query Analysis – the process by which a search query is analyzed to optimize it to bring in the best results.

R

R – open-source software environment that is used for statistical computing.

Radio Frequency Identification (RFID) – technology that makes use of wireless communication to send information about a specific object from one point to another point.

Real Time – descriptor for data streams, events, and processes that have actions performed on them as soon as they occur.

Recommendation Engine – an algorithm that is used to analyze purchases by customers and their actions on a specific e-commerce site; the data is used to recommend other products, including complementary ones.

Records Management – the process of managing a business's records though their whole lifecycle – from the date of creation to the date of disposal.

Reference Data –data that describes a particular object and its properties. This object can be a physical or virtual.

Report – information gained from a dataset query and presented in a predetermined format.

Risk Analysis – the use of statistical methods on at least one dataset to determine the risk value of a decision, project, or action.

Root-Cause Analysis – the process by which the main cause of a problem or event is determined.

S

Scalability – ability of a process or a system to maintain acceptable levels in performance as scope and/or workload increases.

Schema – Defining structure of data organization in a database system.

Search – the process that uses a search tool to find specific content or data.

Search Data – the aggregation of data about specified search times over a specified time period.

Semi Structured Data – data that has not been structured with the use of a formal data model but provides alternative means of describing the hierarchies and data.

Server – virtual or physical computer that serves software application requests and sends them over the network.

Solid State Drive (SSD) – sometimes called a solid state. It is a device that persistently stores data by using memory ICs.

Software as a Service (SaaS) – application software used by a web browser or thin client over the web.

Storage – any means that can be used to persistently store data.

Structured Data – data that is organized with the use of a predetermined structure.

Structured Query Language (SQL) – programming language that is specifically designed to manage data and retrieve it from a relational database.

T

Terabyte – 1000 gigabytes

Text Analytics – application of linguistic, statistical. and machine learning techniques on sources that are text-based to try to derive insight or meaning.

Transactional Data – data with an unpredictable nature, i.e., accounts receivables data, accounts payable data, or data that relates to product shipments.

U

Unstructured Data – data that does not have any identifiable structure, i.e., emails or text messages.

V

Variable Pricing – the practice of changing a price on the fly as a response to supply and demand. Consumption and supply have to be monitored in real time.

Conclusion

I am sure that by now, you would have realized the importance of having a sound system in place to manage your data. In order to manage that data effectively, you might need to expand your organization to include people that are skilled in analyzing and interpreting all that information. With effective data management, you will find it much easier to be analyzed.

With increasing competition, predictive analytics is also gaining more and more importance over time. I have discussed several case studies of large organizations that are using data to expand and improve their operations further. I hope that the information given in this book provided you with insight into the field of predictive analytics.

Through the use of big data analysis, which I have covered extensively in multiple chapters of this book, you are able to see how various industries – from gaming to agriculture – are able to increase their revenues, keep customers satisfied, and increase their final product yield.

I have also touched on the potential dangers and pitfalls of big data, such as the dangers of privacy intrusion and the possibility of failure in business intelligence projects. While these dangers are only a part of the equation, they are certainly something that will be a major part of the big data game in the years to come and should thus be worth keeping a very careful track of. While I firmly believe big data is the future of business, if these things are not considered by the business community, it may be too late to do so at a later point.

I hope you enjoyed this book and apply the techniques discussed to your business.

Lastly, I'd like to ask you a favor. If you found this book helpful or if you enjoyed this book, then I'd really appreciate it if you leave a review and your feedback on. I'd love to hear from you!

Thank you again.

Daniel Covington

CPSIA information can be obtained
at www.ICGtesting.com
Printed in the USA
LVOW04s0023140816
500277LV00019B/1156/P